FARM MACHINERY

Third Edition

BRIAN BELL

Farming Press

First published 1979
Third edition 1989

British Library Cataloguing in Publication Data

Bell, Brian, *1935–*
    Farm Machinery – 3rd. ed.
    1. Great Britain. Arable farming machinery
    I. Title
    631.3'0941

    ISBN 0-85236-199-8

Published by Farming Press Books
4 Friars Courtyard, 30–32 Princes Street
Ipswich IP1 1RJ, United Kingdom

Distributed in North America
by Diamond Farm Enterprises
Box 537, Alexandria Bay, NY 13607, USA

Phototypeset by Galleon Photosetting, Ipswich
Printed and bound in Great Britain by Redwood Burn Ltd, Trowbridge, Wiltshire

CONTENTS

Preface to the Third Edition

Tractor and farm machinery design has made great strides even since the last edition of *Farm Machinery* was prepared in 1985. In recent years, four-wheel-drive tractors have become commonplace on arable farms. The average engine power of new tractors sold to farmers is in excess of 60 kW (80 hp). Tractor cabs provide the driver with greater luxury than that found in many family cars. Top of the range models have computerised monitoring systems to help the driver attain maximum performance from both tractor and implement.

Large field equipment such as balers, forage machinery and sugar beet harvesters has performance monitoring systems with a digital display in the tractor cab. Combine harvester technology has made similar progress with performance and grain loss monitoring systems installed in air-conditioned cabs. The combine harvester pictured on the front cover has a fully computerised information and monitoring system. The entire operator's manual is stored in the computer memory and can be consulted at the touch of a button.

Such developments have led to substantial changes in the Third Edition. In its new format, *Farm Machinery* has much fresh material, including 150 new plates in a total of 270 illustrations. The variety of farm equipment is vast and, even within one group of machines, there are many principles to understand. This book is intended to help machinery users of all ages, whether they are students on formal courses or simply interested in the equipment they operate, to increase their working knowledge of today's 'high tech' farm equipment.

Brian Bell

Introduction

This book is designed to give the reader a sound knowledge of the wide range of tractors and farm machinery in current use. There are sections on tractors, cultivation and drilling equipment, and also crop treatment and harvesting machinery. The final part of the book deals with farmyard and estate maintenance machinery, farm power and the farm workshop. Readers interested in learning more about workshop equipment and processes in the workshop will find much detailed information in *Farm Workshop*, also written by Brian Bell. Other companion books are *Farm Crops* and *Farm Livestock*, both written by Graham Boatfield.

Farm Machinery will prove an invaluable guide to students taking both full-time and part-time college courses in preparation for the various levels of the new National Vocational Qualification (NVQ) awards in agriculture and farm machinery. The book will also prove useful to those embarked on an Open Learning course or for the private study of anyone interested in learning more about modern tractors and farm equipment.

It is not possible to give specific information about any particular tractor or machine in a book of this type. Average figures are given to illustrate typical dimensions and settings. However, the tractor or machinery operator manual supplied by the manufacturer should be considered compulsory reading for anyone using farm equipment.

References to Farm Safety Regulations summarise their main requirements. The reader must study the relevant information leaflets, and in particular those concerning the safe use of pesticides, to find the precise details of the many regulations.

Dimensions and settings are given in metric units (with some approximate Imperial equivalents); a list of metric conversions can be found on page 257.

Acknowledgements

I am grateful to many individuals and companies involved in the manufacture of tractors and farm machinery who have helped in the revision of this book. Numerous illustrations have been provided by these companies; their source is acknowledged in the caption of the plate or figure concerned. Special thanks are due to Ray Broadhurst, John Briscoe, Roger Cutting and Tim Fogden, together with colleagues at Otley College.

FARM MACHINERY

Chapter 1

FARM TRACTORS

Tractors have been used on British farms since the start of the twentieth century. By 1920, petrol-engined tractors were beginning to replace the horse. They were used to pull implements from a drawbar and to drive stationary equipment, such as a threshing machine, by means of a belt pulley.

By 1940, tractor design had progressed to the stage where a hydraulic lift system had been developed and the power take-off shaft was available to drive trailed implements. During the immediate post-war years, the tractor population exploded. Design improvements were numerous: diesel engines, the differential lock, four-wheel-drive and safety cabs are a few examples. The very latest tractors have in-cab computers

PLATE 1.1 *A 135 kW six cylinder diesel engined tractor with a computer based electronic control and monitoring system in the cab.* (Massey-Ferguson)

which allow the driver to keep a check on many operational procedures from the comfort of an air-conditioned cab.

Most of the early tractors had an engine power in the region of 15–22 kilowatts (20–30 hp). At the present time, tractor engine power on medium size farms will usually be up to about 75 kW (100 hp) and frequently up to 120 kW (160 hp) on larger holdings. Some of the very large arable estates have much bigger tractors with engine power of 150 kW (200 hp) or more.

Narrow country roads sometimes restrict the size of tractors and implements, especially on mixed farms away from the major cereal growing areas of the country.

WHAT THE TRACTOR CAN DO

Pulling

From the very early days of farm mechanisation, tractors have been used to pull a variety of trailed implements. Many of the larger machines

such as drills, balers, forage harvesters and trailers are towed from the drawbar. A swinging drawbar which provides both height adjustment and sideways movement is needed for the heavier trailed machines. Some small tractors have a drawbar attached to the lower hydraulic links; this may only be used for light, trailed implements.

Most tractors have an automatic pick-up hitch. Operated by the hydraulic system, it enables the driver to attach trailed implements without leaving the seat. A hook is lowered and guided under a ring at the end of the implement drawbar. The hook is lifted hydraulically, then locked in the raised position. A big advantage of the pick-up hitch is that much of the implement weight is carried directly by the tractor, giving better wheel grip and improved performance.

Implements are also pulled when attached to the three point linkage of the hydraulic system. This method of implement attachment transfers much of the implement weight on to the tractor. Some of the larger mounted implements have wheels at the rear and are carried on the lower

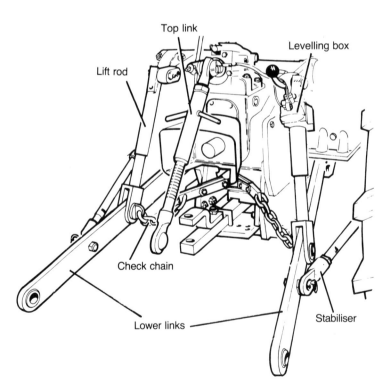

FIGURE 1.1 Swinging drawbar and three point linkage.

hydraulic links with no top link connection. This system of attachment is known as semi-mounted: the front end is lifted and lowered by the hydraulic linkage, while an external ram linked to the rear wheels lifts and lowers the back of the implement.

Lifting

As well as lifting and lowering mounted implements, the hydraulic system is used to operate manure loading equipment, tipping trailers, tractor mounted fork lifts and many other machines. External hydraulic rams provide the power to do this work. They are connected by flexible hoses to the main hydraulic system and its controls. Some further examples of the use of external rams are lifting drill coulters from work, setting the cutting height of a forage harvester and adjusting the cutting position of a hedge trimmer.

Driving

The power take-off is the driving force for a vast array of farm machines such as power cultivators, balers, manure spreaders, sprayers and irrigation pumps. It also acts as a standby on dairy farms to drive the vacuum pump if the electricity supply fails.

The power take-off shaft is driven from the gearbox and has separate clutch control which allows the shaft to continue driving when the transmission clutch is used to stop the forward, or reverse, movement of the tractor. There are two standard speeds for the operation of the power take-off shaft, 540 rpm and 1,000 rpm.

Another function of the hydraulic system is to drive implement mechanisms by means of an hydraulic motor. Some fertiliser spreaders, hedge trimmers and slurry tankers, for example, are driven by hydraulic motors. Oil from the hydraulic system is pumped to the motor through flexible pipes and then returned to the tractor. The control mechanism gives the driver step-less motor speeds in both forward and reverse direction.

The tractor battery is used to provide the power needed by some implement systems. Precision seeders have an electrical monitoring unit with indicator lights in the cab to warn the driver of faults in the seeding mechanism. Electricity powers the remote control systems needed on

some implements. Many crop sprayers have electrically operated taps which turn the spraybar off or on, while the delivery chute angle of some forage harvesters is adjusted by an electric motor.

There are still a few tractors in use with a belt pulley. This was a very important method of driving stationary equipment before the widespread use of electric motors. It is still possible to find farms where saw benches and other fixed equipment is driven by a belt pulley and heavy flat belt. The pulley was attached to and driven by the tractor gearbox. Some pulleys were fitted to the power take-off shaft.

Starting the Tractor

The driver must always be in the tractor seat before starting the engine. Tractors have a safety start mechanism requiring either the gear levers to be in neutral or the clutch pedal to be pressed fully down before the starter can be operated. Early models of tractor may not have a safety start switch, so extra care must be taken to ensure these tractors are started safely.

To start a diesel-engined tractor:

1. Check that the gearbox is set in neutral, especially the lever connected to the safety switch. It is a good idea to push the clutch pedal down too, as this reduces the load on the starter motor. The clutch pedal must be pushed down on tractors which have a pedal-operated safety switch.
2. Set the throttle between half and three-quarters open. Turbocharged engines must not be allowed to run at more than half throttle for at least two minutes after starting from cold.
3. Set the stop control in the run position.
4. Turn on the ignition switch.
5. Operate the starter switch, usually a second movement of the ignition key. If the engine fails to start after thirty seconds, release the starter switch, allow the starter motor to come to rest, then try again after a few seconds. Excessive use of the starter without rest can damage the battery.

Most tractors have some form of starting aid for use in cold weather. A heater unit or excess fuel device on the injection pump are common cold start aids. Failure to use the starting aid in cold weather may well result in a discharged battery.

To stop the engine
Pull the stop control knob out. This will cut off the fuel supply from the injection pump to the engine. *Note*: There are some David Brown tractors which are stopped by pushing the stop control knob in.

The ignition switch is used to stop the diesel engine on some models of fork lift truck. The switch operates a solenoid which cuts off the fuel supply.

When the engine has stopped, remember to set the gear lever in neutral or park and apply the handbrake.

THE DRIVING CONTROLS

Small tractors have a basic set of instruments and controls; the larger and more expensive tractors, however, have many more and some have the luxury of a computerised instrument panel and an air-conditioned cab.

Throttle Used to set the engine speed. Most tractors have both hand and foot throttles. The hand throttle is used to set engine speed for continuous work such as ploughing; this speed is maintained by the governor. The foot throttle can over-ride the hand throttle and is used when frequent changes in engine speed are required for such work as loading manure or driving on the public highway.

Clutch pedal Drive from the engine to the gearbox is engaged and disengaged by the transmission clutch. Some tractors have a dual clutch: the pedal is pushed half-way down to stop the wheels and fully down to disconnect drive to the power take-off shaft. Tractors not fitted with a dual clutch have an independent clutch to control the power take-off. The clutch pedal is on the left-hand side. It must not be used as a foot rest when driving as this will cause undue wear of the clutch thrust bearing.

Brakes Tractor brakes are used:

1. To stop the tractor.
2. To park.

PLATE 1.2 *The view from the driving seat. The gear levers are in the centre, the hydraulic controls to the right and the handbrake on the left. A full range of instruments is found on the panel in front of the steering wheel.* (Massey-Ferguson)

3. To assist with steering, especially on head-lands and in confined spaces.

There are two brake pedals, one for each rear wheel. They are used independently on the field to assist steering and locked together for road work. It is very dangerous to drive a tractor on the road without first locking the brake pedals together. The handbrake is for parking only. Where there is no handbrake, the pedals have a latch mechanism which is used to lock the brakes on when parking.

Differential lock The differential, which allows the driving wheels to turn at different speeds, cannot operate when the diff-lock is engaged. This is usually done with a foot pedal but some tractors have a semi-automatic system of engagement and disengagement. The rear wheels, and the front wheels on a four-wheel-drive, are unable to turn at different speeds with the diff-lock engaged. This reduces wheelspin and increases traction when working in difficult soil conditions.

The diff-lock is designed to disengage automatically but can be released by applying one brake pedal or pushing the clutch pedal down if it fails to disengage when required.

Gear levers Used to select the required forward gear, reverse gear or neutral. Most tractors have two levers, one to select either high or low range and the other to choose a suitable gear ratio for the work in hand. A typical tractor will have eight forward and four reverse gears. Many tractors have an additional high and low range system. Operated electrically with a switch or lever, it can provide a selection of at least sixteen forward and eight reverse speeds. Change on the move gearboxes are used on most tractors of recent manufacture, making gear changing as simple as that for a motor car.

Power take-off Drive is engaged and disengaged with a lever, without the need to use the transmission clutch. Tractors with a dual clutch system need to have the clutch pedal pushed fully down before using the power shaft lever.

Many tractors have a dual speed power shaft with standard speeds of 540 and 1,000 rpm. A separate lever is used to select the required shaft speed. The 540 rpm shaft has six splines; the larger 1,000 rpm shaft has twenty-one splines and it is used to drive implements with a high power requirement.

Hydraulics There are numerous hydraulic controls and each manufacturer has special features in the design of the hydraulic system. The basic controls are:

1. Draft control, used for implements which work in the soil such as a plough or cultivator. This system maintains a regular working depth without the use of depth wheels, provided that soil conditions are reasonably constant.
2. Position control is used for implements which need to be held on the linkage at a constant height above the ground. Hedge trimmers and sprayers are examples of the use of position control.

3. Remote control. Many implements have hydraulic rams operated by the tractor hydraulic system. These external rams are controlled by one or more levers, depending upon the number of rams in use. Tipping trailers, potato harvesters and reversible ploughs are just three of the many machines which have external rams.

The tractor instruction book should be carefully studied to learn the correct operation of the hydraulic controls to ensure maximum efficiency.

The Instrument Panel

There is tremendous variation in the number of dials, gauges, switches and warning lights on the many makes and models of tractors used on our farms. Here again the tractor instruction book should be studied to find out about the instruments on your tractor.

Proof meter This records the number of hours worked by the engine; it also shows the engine speed in rpm. Most tractors record the hours worked by the engine when running at about two-thirds maximum speed. At full throttle, more hours will be recorded than shown by the driver's watch. Some of the latest models, however, record the actual hours the engine runs regardless of engine speed.

The forward speed in each gear is shown by some proof meters, but most have a chart on the instrument panel giving the forward speed in each gear at stated engine speeds.

The throttle setting which gives the two standard power take-off speeds is also marked on the proof meter.

Gauges The number and purpose of gauges clustered on the instrument panel will vary with model of tractor. Engine oil pressure, engine coolant temperature, an ammeter to register rate of battery charge and discharge and a fuel gauge are the most common. Low oil pressure and battery discharge are often indicated by a coloured warning light instead of a gauge.

Indicator lights There are a number of indicator lights. As well as low oil pressure and battery discharge, lights are used to warn of airflow

PLATE 1.3   *This electronic instrument panel provides the driver with a wealth of information. The top section has a number of engine and electrical system warning lights. A hare (illuminated) and a tortoise (not shown) indicate selection of high or low speed ratio. The left centre panel has bargraphs showing fuel level, coolant temperature, oil pressure and battery condition. The centre panel bargraph shows engine speed (1,600 rpm). Three touch buttons are used to select readings for engine, forward and power take-off speed; the reading shows a forward speed of 14.3 km/h. Total engine hours are also shown.*

The right centre panel offers information, at the touch of a button, on area worked, area worked per hour, time worked and wheelslip. The read-out indicates an area of 46 hectares has been completed. Three light switches are situated at the bottom left corner and the four-wheel-drive engagement switch is covered by the driver's hand.   (Ford)

PLATE 1.4 *A tractor cab equipped with a computerised performance monitor* (top) *and an electronic linkage control system* (black panel). *Three levers beside the seat control auxiliary hydraulics and the lever near the right-hand arm rest is for the power take-off shaft.* (Massey-Ferguson)

restriction, low fuel level, parking brake on, traffic indicators and hazard lights, low hydraulic oil pressure, rear working lamps on, main beam on, etc.

Electronic instrument panels Some tractors have electronic instruments; a typical example has eighteen coloured indicator lights. They monitor ignition, parking brake, low fuel, main beam, air cleaner restriction, engine coolant temperature, engine oil pressure, indicators, rear working light, etc.

Bar graphs replace gauges to indicate engine speed, engine oil pressure, engine coolant temperature and fuel level.

The panel is completed by a set of switches for lights, traffic indicators, four-wheel-drive engagement, automatic high and low gear ratio, starter key, air conditioning, horn button and radio controls.

In-cab computers Available on some tractors in the form of an electronic performance monitor which provides information at the touch of

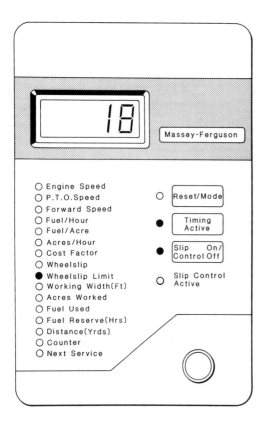

PLATE 1.5 *Detail of the performance monitor panel. The knob* (bottom right) *is used to select from the menu on the left-hand side. Wheelslip limit has been selected at setting 18. If wheelslip exceeds this level, the implement is automatically lifted to reduce the amount of slip. To find when the next service is due, the bottom item is selected and the digital read-out will give this information.*   (Massey-Ferguson)

a button. A digital read-out will give engine, power take-off and forward speeds, fuel used per hour and per acre, fuel reserve, wheelslip details, total engine hours and when the next service is due.

Computerised controls can also automatically disengage the diff-lock when the hydraulic linkage is raised, re-engaging it when the linkage is lowered. The system also engages four-wheel-drive when braking and when the tractor is parked. On one model, when it travels at speeds in excess of 14.5 km/h (9 mph), the four-wheel-drive automatically disengages to reduce tyre wear and fuel consumption.

TYPES OF TRACTOR

The size of four-wheel tractors varies from the small ride-on model used by gardeners and smallholders up to the giant tractors found on the very large arable farms. Tractor power on livestock farms will be around 30–37 kW (40–50 hp) while predominantly arable farm tractors are in the 50–75 kW (70–100 hp) range or more.

Rowcrop Tractors

Mainly used by market gardeners and farmers who grow root and vegetable crops, this type of tractor has a small turning circle, easily adjusted wheel track setting and offers good visibility of the growing crop. A variation of the rowcrop tractor is the self-propelled tool carrier. It has a tool frame between the front and rear wheels which can carry implements for rowcrop work. Implement attachment is simple and the tool carrier offers very good visibility of the crop.

General Purpose Tractors

As the name suggests, they do most of the work on both arable and livestock farms. Engine size will vary from 45–75kW (60–100 hp) or more. General purpose tractors have heavy duty, live hydraulics and live power take-off shaft. The larger models will also have a two-speed power shaft. These tractors are capable of pulling heavy loads from the pick-up hitch, drawbar or hydraulic linkage.

Many general purpose tractors will also have four-wheel-drive: the front wheels will be smaller than the rear and the driver has the option to engage four-wheel-drive when the soil conditions or load require extra power.

Tracklayers

Mainly used by contractors and farmers who have large, heavy land farms. A typical farmer-owned tracklayer has a 65 kW (85 hp) engine.

PLATE 1.6 *A compact tractor, used for horticultural and light farm duties including work around livestock buildings. The model illustrated has a 16 kW (22.7 hp) engine. Four-wheel-drive and hydrostatic transmission systems are available for many models of compact tractor.* (Ford)

Larger models with engine power of 110 kW (150 hp) or more are mainly used by contractors for earth moving and drainage work.

Tracklayers may only be used on the roads after street plates have been fitted to the tracks; these prevent damage to the road surface. Most tracklayer owners use a low loading trailer to transport the machine from one site to another.

Although tracklayers have rather low operating speeds, they can do good work in conditions where wheeled tractors would not be able to work efficiently. The tracks exert a very low ground pressure compared with wheels, which means that soil compaction is reduced to a minimum.

Four-wheel-drive Tractors

There are two main types of four-wheel-drive tractor. One has four large equal size wheels, the other has large rear wheels and smaller diameter front wheels.

Most four-wheel-drive tractors are alternative versions of two-wheel-drive models. The front wheels have traction tyres and drive is directly from the gearbox with a differential in the front axle unit. On some tractors four-wheel-drive is engaged with a lever, while others have an automatic drive engagement system which brings front wheel drive into operation when conditions require extra power.

PLATE 1.7   *An example of a rowcrop tractor with two-wheel-drive and a 40 kW (54 hp) engine. Easily adjusted wheel track settings and good visibility are important features of a rowcrop tractor.*   (Ford)

Four-wheel-drive with equal size wheels is less common. Most models are special conversions of standard production tractors with a power range of 75–110 kW (100–150 hp).

All four-wheel-drive tractors have the advantage of increased pulling power, especially in difficult conditions, compared with similar powered two-wheel-drive versions. They are ideal for heavy work such as ploughing, cultivating and root harvesting.

Articulated Four-wheel-drive Tractors

There are a few articulated four-wheel-drive tractors in Great Britain. Brought to this country from America, these huge tractors which pivot between the front and rear wheels (i.e. articulate) are very expensive. They offer high working speeds and can handle very wide implements. Engine power ranges from 150–260 kW (200–350 hp). Typical fuel tank capacity is around 1,125 litres.

PLATE 1.8 *A tracklaying tractor which has a rather low forward speed but causes very little soil compaction. The main disadvantage of tracklayers is the problem of moving them along the public highway.* (Caterpillar Inc)

Power is transmitted to the ground by four pairs of large dual wheels. The hydraulic lift takes category three implements and the 1,000 rpm power shaft is able to deliver very high horse power.

Using Tractors On The Road

The Highway Code must be obeyed whenever a tractor is driven on the public highway. The maximum legal speed for driving a tractor on the road is 32 km/h (20 mph). This is reduced to 19 km/h (12 mph) when towing two trailers or driving a wide self-propelled machine such as a combine harvester.

When a tractor and trailer is driven on the road, the driver must remember these points:

- Make sure the tractor brakes are in good working order. The pedals must be latched together.

PLATE 1.9 *A tracklaying tractor with rubber tracks. It has a top speed of nearly 30 km/h and can be used on the road in the same way as a wheeled tractor.* (Caterpillar Inc)

PLATE 1.10   *A general purpose four-wheel-drive tractor.*   (Case IH)

- The windscreen must be clean; make sure the screen washer and wiper work properly.
- Suitable mirrors must be fitted to the tractor to give the driver a clear view of the traffic behind, even when pulling a loaded trailer.
- The steering and tyres must be in good condition.
- Lights, when fitted, must be in working order at all times.
- A trailer must have effective brakes which can be applied progressively from the tractor seat.
- Amber flashing road beacons must be used when driving on dual carriageway road with vehicles having a maximum speed of less than 40 km/h (25 mph). Flashing beacons may be used on other roads to warn traffic of slow moving vehicles.
- Special regulations apply when taking wide or projecting loads on the highway. If tractor lights and reflectors are obscured, there must be additional lights on the load. Marker plates are required for long mounted or semi-mounted implements as well as wide or overhanging loads.

Road Traffic Regulations are very complex: the information above is merely a guide. Tractor and self-propelled machinery drivers must find out the exact requirements for their equipment to ensure they are within the law.

Other points to remember when using a tractor or self-propelled machine on the road include:

- The tractor must have a road tax disc or a tax exemption certificate. There must be a valid insurance certificate and the registration plate must be clearly visible from the rear. The driver must have a licence too!
- When towing a laden trailer, make sure the load is secure.
- Do not allow mud or other material to fall on the road. Where this cannot be avoided,

PLATE 1.11 *This articulated four-wheel-drive tractor pivots between the front and back wheels. With its eight wheels, this giant tractor has a 195 kW (260 hp) engine.* (Massey-Ferguson)

the law requires that it must be removed to avoid creating a hazard to other road users.

In addition to the many Road Traffic Regulations, there are requirements listed in the Agricultural Field Machinery Regulations, Avoidance of Accidents to Children Regulations and the Health and Safety at Work Act which must be observed. You should read the official leaflets to find the exact terms of the various regulations. The following notes are only a guide:

- Children under thirteen must not ride on or drive a tractor or farm machine. They may ride on an empty, or partly filled trailer provided that all four sides project above the floor or load.

In at least one county (Norfolk) the minimum age for driving a tractor solely on the farm is fourteen years.

- Passengers must not ride on the drawbar or other linkage of a tractor or any other machine when towing is taking place.
- A tractor or other self-propelled machine may only be started or set in motion when the driver is in the normal driving position.
- The driver must not leave the seat while the tractor is in motion, except in an emergency.
- Drawbar pins must be secured with a safety clip to prevent them jumping out while towing.
- All cutter bars must be covered with a strong, rigid guard when not in use and when taken on the road.

Children are killed in farm accidents every

year. Always take extra care when children are playing nearby. When you need to back a trailer, check there is no one behind before reversing.

Quiet cabs isolate the driver from the sound of children playing, also from the noise made by other workers and their equipment. Take special care when driving around farm buildings and be prepared for an unexpected hazard just around the corner.

Safe Driving in the Field

- Take special care when driving on sloping land. Select a low gear before driving downhill and never change gear except when the tractor is stationary.
- Make sure that nobody is in the way when moving off, lowering an implement or engaging the drive to a machine.
- Do not drive close to the edge of ditches or dykes. Stay clear of the edge of silage clamps when consolidating them with a tractor.
- Drive at a safe speed at all times. Good drivers do not need to apply the brakes fiercely as they are in full control of the tractor.

Safe Parking

- Always apply the handbrake when parking a tractor. The gear lever should be left in neutral or park and the ignition key should be removed.
- Always lower implements and loaders to the ground when parking a tractor.

Safe Hitching

- Back squarely up to mounted implements when preparing to attach the linkage arms. Do not try to manhandle heavy equipment into position.
- Never position yourself between tractor and implement when hitching. Always work from the side. Remember the correct sequence for attaching the linkage is left—right—top.

- When towing from a swinging drawbar do not use double jaws on both tractor and implement. This practice can shear the drawbar pin. The solution is to remove one half of the jaw from the tractor drawbar.

All agricultural workers have a legal responsibility to make full use of guards and other protective devices on tractors and farm equipment. Guards must be in position when farm equipment is in use. Broken or damaged guards must be reported to the employer so that repair or replacement can be arranged. The Farm Safety Regulations also state that the worker has a duty to ensure that his actions, while at work, do not affect the health and safety of others.

There are numerous publications on farm safety matters. You should read them carefully and follow their recommendations.

SUGGESTED STUDENT ACTIVITIES

1. Read your tractor instruction book.
2. Get to know the controls of as many different tractors as possible.
3. Study the Highway Code, taking special note of the sections which deal with driving tractors on the public highway.
4. Learn and obey the Farm Safety Regulations, especially those concerned with tractors and farm machinery.
5. Visit a vintage tractor rally or a farm museum. Look at the tractors and machinery and take note of how design has changed over the years. (Vintage tractors are twenty-five years old or more.)

SAFETY CHECK

The law requires the driver to start the engine of a tractor or self-propelled machine only from the driving seat. The driver may not leave the driving position while the machine is in motion—except in an emergency.

Chapter 2

TRACTOR ENGINES

Farm tractors are powered by diesel engines, which work on the four-stroke principle. This type of engine is called an internal combustion engine because the heat required to make it work is produced inside the engine cylinders. Steam engines are called external combustion engines because the heat is created by burning coal to produce steam, which is passed to the cylinder.

Petrol engines are also in use on many farms. A few petrol-engined tractors can be seen working in marginal farming areas. Small petrol engines are used to drive some pumps and similar small stationary equipment. Chain saws have two-stroke petrol engines.

THE FOUR-STROKE ENGINE

A four-stroke engine has one working or power stroke in every four. A two-stroke engine has

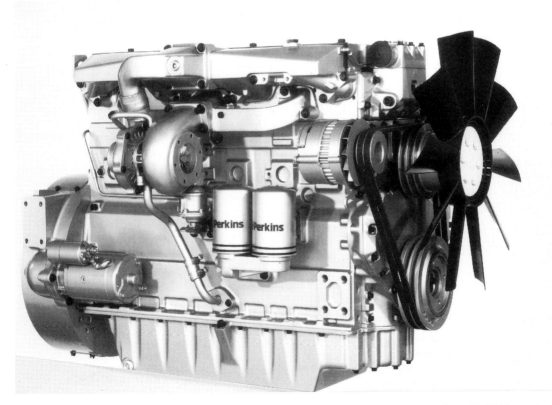

PLATE 2.1   *Turbocharged six-cylinder diesel engine. Versions of this engine produce from 70–195 hp.*   (Perkins)

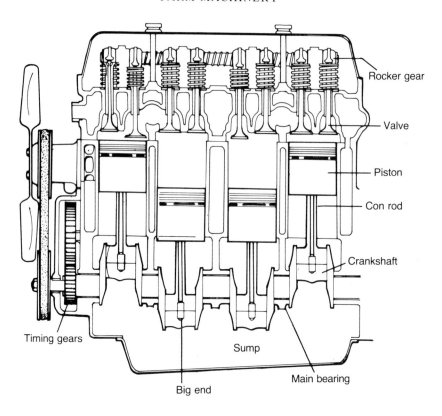

FIGURE 2.1   Section of a four-stroke engine.

every second stroke as a power stroke. The basic principle of a four-stroke engine is that by burning a mixture of fuel and air above a piston in a cylinder, heat is produced. Expansion, caused by the heat, forces the piston downwards and turns a crankshaft.

Tractors, with few exceptions, have three-, four- or six-cylinder engines. Each cylinder has two valves. The inlet valve allows air into the cylinder of a diesel engine. It lets fuel and air into the cylinder of a petrol engine. The exhaust valve releases waste gases to the exhaust pipe. The valves are opened by pushrods and rocker arms which are operated by the camshaft. This is driven by and timed with the crankshaft, to open the valves at the correct time during the four-stroke cycle. The valves are closed with strong springs.

Most multi-cylinder engines have overhead valves. This means that the valves are in the cylinder head, above the pistons. Small, single-cylinder engines usually have side valves. The valves are in the cylinder block, close to the pistons and move upwards when they open. Overhead valves are more efficient than side valves but the simple design of the side valve gear makes it ideal for small petrol engines.

The piston is joined to the connecting rod (con rod) with a gudgeon pin, which runs in the little end bearing. The con rod is held on the crankshaft by the big end bearing. Each piston has a set of piston rings. The top rings are the compression rings. They act as a seal between piston and cylinder wall to prevent loss of pressure above the piston on the compression stroke. The bottom ring is an oil ring. Some engines have two oil rings. Their purpose is to scrape oil from the cylinder walls to prevent it getting above the piston, into the combustion chamber.

The crankshaft is secured in the cylinder block by the main bearings. A four-cylinder diesel engine has five main bearings. A heavy fly-

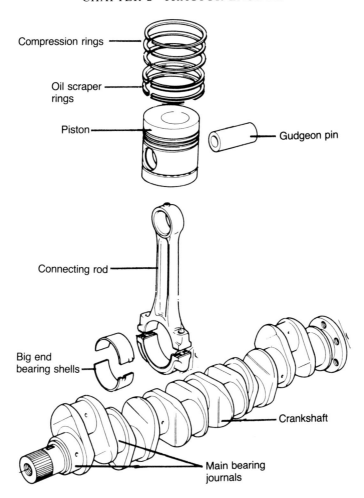

Compression rings

Oil scraper rings

Piston

Gudgeon pin

Connecting rod

Big end bearing shells

Crankshaft

Main bearing journals

FIGURE 2.2   Piston and crankshaft components.

wheel is bolted to the gearbox end of the crank-shaft. It stores energy and keeps the crankshaft running smoothly between the power strokes. The timing gear, at the front of the crankshaft, drives the camshaft and, on a diesel engine, the injection pump. The camshaft is timed to open the valves at the correct point in the four-stroke cycle. The injection pump is timed to inject fuel at the correct time on the compression stroke. Some timing systems are chain driven from the crankshaft. The timing of the spark on a petrol engine is by a gear on the camshaft which drives the distributor.

All internal combustion engines have a cooling system to remove unwanted engine heat. Water-

and air-cooling systems are used. Lubrication is vital. All engines have their own lubrication system to ensure a constant flow of oil to all moving parts.

The air cleaner is another essential part of an engine. It provides a constant supply of dust-free air to the cylinders. Spark ignition engines have a carburettor and fuel system to supply petrol. The ignition system provides an electric spark which ignites the mixture of fuel and air.

Compression ignition engines have a fuel system to provide exact amounts of fuel to the cylinders. The following pages illustrate and deal with the components and systems listed above.

The Four-stroke Diesel Engine

The diesel engine relies on immense heat inside the cylinder to burn the fuel as it is injected into the engine. The high temperature is achieved by compressing the air in the cylinders to a very high pressure, approximately 34 bar (500 psi); this gives a temperature in the region of 550°C. Because the fuel is ignited by heat, created through compression, the diesel engine is also known as a compression ignition engine.

The four-stroke diesel works in this way:

Induction stroke   The piston travels down and air is sucked from the air cleaner into the cylinder through the open inlet valve.

Compression stroke   Both valves are closed. The rising piston compresses the air in the cylinder to a very high temperature. Just before the piston reaches the top of the stroke, a fine spray of diesel fuel is injected into the cylinder, above the piston. The fuel burns instantly and the heat causes a rapid expansion of the gases above the piston.

Power stroke   The expanding gases drive the piston down, turning the crankshaft. This is the working stroke of the four-stroke cycle.

Exhaust stroke   The exhaust valve opens at the end of the power stroke. The waste gases are forced out of the cylinder by the rising piston. Just before the top of the exhaust stroke, the inlet valve opens, ready for the next induction stroke.

A tractor engine runs at speeds well above 2,000 rpm. Even at this speed, each piston will complete 4,000 strokes every minute. Each valve will open and close 1,000 times and each cylinder will receive 1,000 separate injections of diesel fuel.

The Four-stroke Petrol Engine

The four-stroke petrol engine has lower working pressures and temperatures than a diesel engine. The fuel is ignited with an electric spark after a mixture of fuel and air has been compressed in the cylinder. Petrol engines are also known

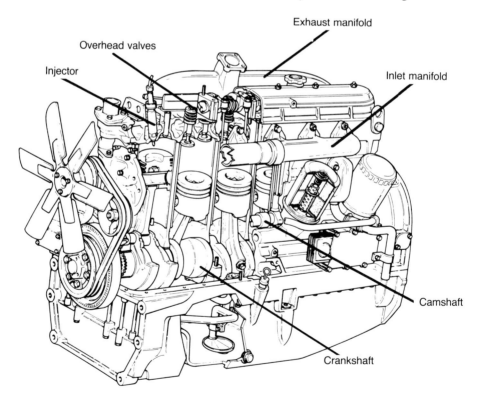

FIGURE 2.3   Sectional view of overhead valve engine. (*Fiatagri*)

as spark ignition engines. A four-stroke petrol engine works in this way:

Induction stroke A mixture of fuel and clean air is sucked into the cylinder through the open inlet valve.

Compression stroke Both valves are closed. The rising piston compresses the mixture of fuel and air in the cylinder. Just before the top of the compression stroke, a spark ignites the mixture.

Power stroke The expanding gases drive the piston down, turning the crankshaft.

Exhaust stroke The rising piston forces the waste gases out of the cylinder through the open exhaust valve. The cycle starts again with another induction stroke.

Firing order

Most tractors have three, four or six cylinders in line. There are exceptions. The firing order, sometimes stamped on the cylinder block, is the order in which the cylinders receive their spark or fuel injection. Number one cylinder is always at the front. The firing order for three-, four- and six-cylinder engines are:

3 cylinder   1.2.3.
4 cylinder   1.3.4.2.   or   1.2.4.3.
6 cylinder   1.5.3.6.2.4.   or   1.4.2.6.3.5.

Compression ratio

This is the relationship between the volume in the cylinder at top dead centre (TDC) and bottom dead centre (BDC). The piston is at TDC when it is at its highest point in the cylinder. BDC is when the piston is at its lowest point in the cylinder.

A spark ignition engine has a compression ratio of about 7:1. This means that the volume in the cylinder at TDC is one-seventh of the volume at BDC. A typical compression ratio for a diesel engine is 16:1.

THE DIESEL ENGINE

A diesel engine is more robust than a petrol engine. It must withstand the much higher working pressures and temperatures resulting from the high compression ratio. The sturdier design makes the diesel engine more expensive to manufacture but is preferred for tractors for its greater efficiency and economy compared with a spark ignition engine with similar dimensions. The diesel engine develops full power very soon after starting from cold and is relatively trouble free.

Direct and indirect injection

These are the two types of diesel engine used in agriculture. The direct injection engine has the fuel injected directly into the cylinder, above the piston. This is the hottest part of the engine and gives direct injection the advantage of good starting performance, even on cold mornings.

The indirect injection engine has the fuel injected into a pre-combustion chamber at one side of the main combustion space above the piston. This design improves engine efficiency because the pre-combustion chamber ensures the fuel mixes thoroughly with the air as injection takes place. Indirect injection engines can be a problem to start, especially on winter mornings. A cold starting device helps solve this problem.

The Diesel Fuel System

A diesel engine requires a precise amount of clean fuel injected into each cylinder at the correct time, in an atomised form. This is the job of the fuel system.

The fuel flows from the tank, through a tap, to the lift pump. The tap is always left on, to prevent air bubbles forming in the fuel. Air in the fuel system causes misfiring. Any amount of air will stop the engine. It will not re-start until this air has been bled (removed) from the fuel.

The *lift pump*, driven either by the engine camshaft or the injection pump, delivers fuel under slight pressure to the filters and injection pump. Some pumps have a filter to collect dirt and water from the fuel. A hand priming lever on the pump is used to bleed the fuel system. Engines without a lift pump rely on gravity feed to the filters and injection pump.

The *fuel filter*—some tractors have two— removes all traces of dirt and water from the fuel. Many filter units have a water trap to collect any water which may occur through condensation in the fuel tank. The fuel filters must be renewed periodically, usually after 600 hours' service. Each filter unit has a bleed screw.

The *injection pump* is driven by the engine

IN-LINE PUMP
SYSTEM

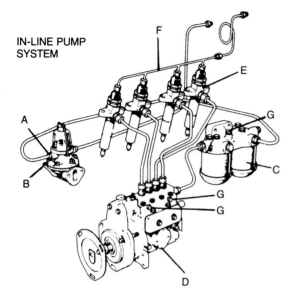

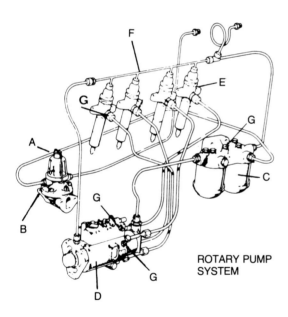

ROTARY PUMP
SYSTEM

A—Lift pump
B—Lift pump priming lever
C—Filters
D—Injection pump
E—Injectors
F—Leak off pipe
G—Bleed screws

FIGURE 2.4   Diesel fuel injection systems.

timing gears. It delivers fuel, in minute quantities, at a pressure of approximately 175 atmospheres (1 atmosphere = 14.7 psi so 175 atmospheres is equal to 2,500 psi). There are two types of injection pump: rotary (or DPA) and in-line.

The in-line pump has a separate pumping element for each injector. It is set by the manufacturer to ensure each cylinder receives an equal quantity of fuel. A worn pump will not be able to do this efficiently and fuel is wasted.

A rotary pump has only one pumping element with a distributing head to supply fuel to each cylinder to suit the firing order. This design ensures that all cylinders receive equal quantities of fuel, even when the pump element is worn.

The throttle setting regulates how much fuel is injected into the cylinders, and in this way, controls engine speed. A governor is built into the injection pump. This maintains the engine speed selected with the throttle. The governor allows the injection pump to supply extra fuel when the engine speed falls, because of an increased load on the drawbar. When the engine races above the selected speed, the governor reduces the fuel supplied to the engine and brings the revs down again.

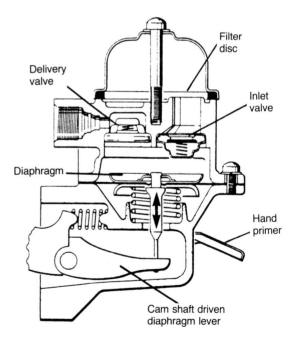

FIGURE 2.5   Fuel lift pump.   (*Perkins Engines*)

The *injector*, sometimes called an atomiser, sprays fuel in a fine mist into the cylinder, when the piston is just a few degrees before TDC on the compression stroke. The holes in the tip of the injectors are very small, often no more than 0.2 mm in diameter. Failure to supply clean fuel to the injectors will cause an increase in the rate of wear. Worn injectors will put too much fuel into the engine. This is a waste of fuel. A sure sign of worn injectors is continuous clouds of black smoke from the exhaust pipe. Injectors should be checked periodically. This is a task for a trained mechanic.

Extra fuel is supplied to the injectors to keep them cool. This is returned to the fuel tank through the leak-off pipes.

CARE OF THE FUEL SYSTEM

Fill the tractor tank at the end of the day's work. This will reduce condensation so that less water contaminates the fuel. Always use clean fuel. Fill the tank through a funnel with a fine mesh filter in the neck if in doubt about the quality of any fuel.

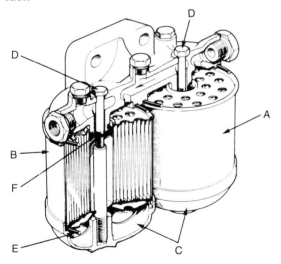

FUEL FILTER

A—Element—first filter.
B—Element—second filter.
C—Bowl.
D—Retaining bolt.
E—Sealing ring—outer.
F—Sealing ring—inner.

FIGURE 2.6 Fuel filter. (*Perkins Engines*)

Some lift pumps have a filter and water trap. Clean this filter when servicing the main filter unit.

Regular servicing of fuel filters is very important because, with use, the filter element becomes partially blocked with dirt and cannot work efficiently. Most diesel engine fuel systems need a new filter element after every 600 hours of running. Always replace the rubber sealing rings at the same time. After servicing, the fuel system will require bleeding to remove all traces of air. Some filter units have a water trap below the element. This should be drained from time to time.

Very few injection pumps need any attention between major overhauls. Some in-line pumps have an oil level plug which should be checked periodically. Check your tractor manual before attempting any maintenance on the injection pump.

The injectors should be checked by a mechanic using an injector tester at intervals of about 600 hours. Faulty injectors can be replaced with exchange units.

Bleeding the fuel system
There will be air in the fuel system after servicing the filters or if the tractor runs out of fuel. The air must be removed by bleeding the fuel system before the engine will start again. To bleed a fuel system:

1. Make sure there is plenty of fuel in the tank. Check that the tap is turned on.
2. Remove the bleed screw on top of the filter unit. Operate the lift pump priming lever until fuel runs from the bleed point free of air bubbles. Tighten the bleed screw. Sometimes it is only necessary to slacken the bleed screw as it is drilled through the centre.

    Engines without a lift pump are bled by gravity. Slacken the bleed screw and wait until the fuel runs free of air bubbles.
3. Bleed the second filter in the same way.
4. Slacken, or remove, the bleed screw on the injection pump and repeat the process. Some pumps have two bleed screws.
5. Engines with rotary pumps usually will not start without first bleeding the injector pipes. Slacken, or remove, at least two of the pipes from the injectors. Run the engine with the starter motor until fuel is

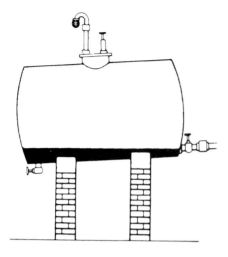

FIGURE 2.7   Diesel fuel storage tank.

seen coming from the pipes. Replace and
tighten the pipe unions. The engine should
now start and will soon run smoothly. See
Figure 2.4 for location of bleed screws.

Cold starting

In cold weather, some diesel engines can be
difficult to start. Indirect injection engines have
some form of cold starting aid. Some direct
injection engines are similarly equipped. The
main types of starting aid are:

- *Excess fuel device*   Extra fuel can be supplied
  to the engine for starting, by pushing in the
  excess fuel button on the injection pump. The
  button returns to its normal position once the
  engine has started.
- *Heaters*   There are two types. Some engines
  have a small heater unit in each cylinder. It is
  often called a glow plug. These heaters warm
  the air inside the cylinders before the starter
  is used. A pilot light on the tractor instrument
  panel shows the driver when the glow plugs are
  hot.

  The second type of heater warms the air in
  the inlet manifold before it enters the cylinders. Extra fuel is sprayed into the heated air
  through a thermostatic valve. The fuel burns
  and, at this moment, the driver operates the
  starter motor.
- *Aerosols*   Containing liquids such as ether
  which burn very rapidly, these can be used

as a last resort in cold weather. The aerosol is
sprayed into the air cleaner or inlet manifold
immediately before starting.

Storage of Diesel Fuel

Clean fuel is essential to ensure long trouble-free
service from diesel injection equipment. The
storage system must keep the fuel as clean as
the oil companies supply it.

The tank should be high enough to allow
gravity filling. It must be sited to give easy access
for tractors and delivery tankers.

The base of the tank should slope away from
the draw-off point. A sludge tap at the lower end
is used to drain off water and sludge, before a
new delivery of fuel arrives.

The tank must not be galvanised, because
diesel fuel attacks the zinc coating.

The tank filler opening should be airtight. A
ventilator pipe at the top of the tank will allow
air in but restrict the entry of dust and water. A
dipstick is needed to check the fuel level.

When a new delivery of fuel is expected, all
tractor tanks should be filled. The sludge should
then be drained off. After delivery, leave the
tank to settle for twenty-four hours before any
fuel is drawn off.

Turbochargers

The power developed by an engine is determined
by the amount of fuel it can burn during the brief
combustion period. There must be plenty of air
to allow the engine to burn fuel. The amount of
air taken into an engine can be increased with a
turbocharger.

A turbocharger is a form of air blower. The
hot exhaust gases drive a turbine (fan) on their
way from the engine to the exhaust pipe. A
second fan, on the same shaft, draws air from
the air cleaner. This fan and its housing form a
compressor unit which supplies compressed air to
the cylinders. This means that more air is forced
into the cylinders and more fuel can be burnt. An
engine which normally develops 75 kW (100 hp)
can have its power increased to about 100 kW
(130 hp) when a turbocharger is fitted.

The turbocharger shaft runs at very high
speeds. When the engine is idling, the turbine
shaft runs at approximately 20,000 rpm and
at full throttle it can reach speeds of around

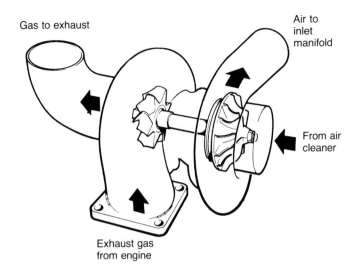

Gas to exhaust

Air to
inlet
manifold

From air
cleaner

Exhaust gas
from engine

FIGURE 2.8 Turbocharger.

80,000 rpm. For this reason proper lubrication is vital. To ensure that a turbocharger is fully lubricated, the engine should be allowed to run at idling speed for approximately one minute before driving the tractor. The engine should also be allowed to idle for a similar period before it is stopped. This will allow the turbocharger and manifold time to cool and therefore prevent possible distortion of the turbocharger.

It is very important to service the air cleaner at frequent intervals on turbocharged engines.

The Intercooler

This cools the air as it leaves the turbocharger before it is allowed to enter the engine cylinders. Cool air is denser than warm air, so by cooling the air as it leaves the turbocharger, it is possible to increase engine performance.

An intercooler consists of an additional fan, driven by the turbocharger itself, which draws cold air over the tube supplying the turbocharged air to the cylinders.

Air Cleaners

Most tractors are now supplied with a dry element air cleaner. However, many of the older tractors still in use have oil bath cleaners.

The dry air cleaner

The engine air supply is sucked through a thick paper, or felt filter element on its way to the cylinders. A gauge on the tractor instrument panel warns the driver if the airflow to the engine is restricted.

A dry air cleaner is cleaned by tapping the element gently against a tractor tyre. This will not harm the filter element, which is useless if damaged. The element should be rotated between each tap on the tyre.

As an alternative, compressed air may be used to clean the element, but the pressure should not exceed 2 bar (30 psi). The air should be used to blow from the inside of the element outwards. Do not hold the air nozzle close to the element as this may damage it.

Some tractor manufacturers advise washing the outer element after 300 hours' service. Use cool water, shake the surplus away and leave the element to dry naturally.

The inner filter element should not be removed. If the air cleaner restriction warning light still glows after servicing the outer element the inner unit should be replaced by a trained mechanic.

In normal working conditions, the dry air cleaner element should be checked weekly and cleaned as necessary. In dusty working con-

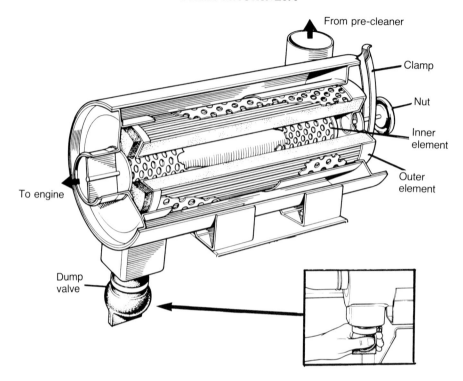

From pre-cleaner

Clamp

Nut

Inner element

Outer element

To engine

Dump valve

FIGURE 2.9   Dry air cleaner.   (*Massey-Ferguson*)

ditions, though, the element should be checked every day.

The dry air cleaner is efficient in most conditions. However, in some countries where dust is a major problem, oil bath air cleaners are used to reduce the risk of serious engine wear.

The oil bath air cleaner

A pre-cleaner, often a dome-shaped cap on the air cleaner inlet, removes large pieces of dirt, such as chaff. The air then passes through the oil bath and oil-impregnated gauze filter. Dust is trapped by the oil in the gauze as the air passes towards the cylinders. When the engine is stopped, the oil collects in the oil bath and the dirt settles out. As the level of dirt rises in the oil bath, the oil level will also rise. The oil level must not be allowed to get too high, so the bowl must be emptied and the dirt removed.

Maintenance is very important. A clogged air cleaner starves the engine of its air supply causing loss of power, waste of fuel and undue wear. In normal conditions check the oil bath every week but in dusty conditions it should be checked daily. When there is about 12 mm of dirt in the oil bath it should be cleaned out and refilled with fresh engine oil to the level mark. Check the gauze filter and clean if necessary. Some filters can be removed from the casing and cleaned, either with an air line or washed with petrol.

THE PETROL ENGINE

Although tractors are no longer made with petrol engines this type of engine is still used on the farm, mainly in its single-cylinder, air-cooled form. It has a carburettor to supply and mix petrol with the air and an ignition system to ignite the mixture of air and fuel. The spark may be produced by a magneto or coil ignition system.

The Fuel System

The fuel flows from the tank through a tap to the carburettor float chamber. A small needle

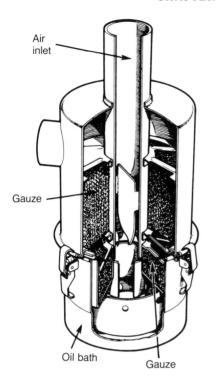

FIGURE 2.10   Oil bath air cleaner.

with an insert which is called the venturi. Because the venturi reduces the size of the air pipe, it also increases the speed of the air as it passes through it. The increase in air speed causes fuel to be drawn from the main jet. The fuel mixes with the air as it is sucked into the cylinders.

The *choke* is an adjustable valve placed between the air cleaner and the carburettor. It is used when starting the engine. The choke restricts the flow of air to the cylinders and in this way supplies an air/fuel mixture with a much higher petrol content. This is called a rich mixture.

The *throttle* controls the quantity of mixture supplied to the engine. The throttle valve is between the carburettor and the cylinders. It is opened and closed with the throttle lever. The governor can over-ride the throttle lever and alter the throttle valve position to maintain the required engine speed.

Air/fuel ratio

For normal running, the carburettor is set to mix one part of fuel with fifteen parts of air by weight. This means that 1 kg of petrol is mixed with 15 kg of air to give an air/fuel ratio of 15:1. The mixture will be much richer, about 8:1, when the choke is in use. The quantities of air and fuel in the mixture can be better understood in terms of volume. An engine with an air/fuel ratio of 15:1 will need approximately 9,000 litres of air for every litre of fuel used.

An engine will not develop full power when running on a weak mixture such as 17:1. A rich

valve, controlled by the float, keeps the fuel in the float chamber at the correct level.

The petrol flows from the float chamber into a small tube (main jet) in the pipe which carries the air to the engine from the air cleaner. The size of the air pipe is reduced around the main jet

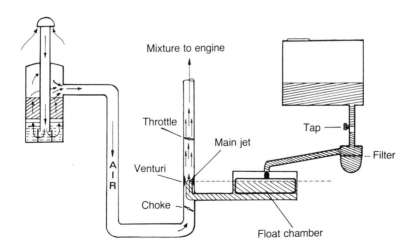

FIGURE 2.11   Petrol engine fuel system.

mixture, e.g. 12:1, wastes petrol and will cause a build-up of carbon around the valves and combustion space. Excess deposits will reduce engine efficiency. Black smoke from the exhaust pipe may be caused by a rich mixture. It can also be due to a blocked air cleaner which has the effect of supplying a rich mixture to the engine. Do not confuse black smoke with blue smoke, which is a sign of the engine using (burning) oil. This happens when the engine is worn. Two-stroke engines also produce blue smoke when too much oil is mixed with the fuel (see page 33).

Care of the fuel system
It is important to use clean fuel and handle it in clean containers. Petrol engines used on farms have a sediment bowl to trap any dirt in the fuel. This should be cleaned periodically. It is usually combined with the fuel tap.

The carburettor float chamber will collect dirt after a time. This should be removed carefully, without damaging the paper gasket, and cleaned with petrol. The carburettor jets are very small, so the fuel system must be kept clean to reduce the risk of blocked jets.

Diaphragm carburettors
Engines used on chain saws and some garden machines often have diaphragm carburettors. One type has a diaphragm instead of the usual float and float chamber. The diaphragm, operated by engine suction, supplies fuel to the

metering chamber. From here, it mixes with the ingoing air as it is drawn into the cylinder. Another design has two diaphragms. One pumps fuel to the metering chamber where the second supplies the fuel in the required quantity to mix with the air as it is drawn into the cylinder.

The Ignition System

The spark which ignites the mixture of fuel and air in the cylinders must be timed to occur just before TDC on the compression stroke of the four-stroke cycle.

Multi-cylinder petrol engines normally have a coil ignition system; this requires a battery to supply the electric current. Single-cylinder engines are more likely to have a magneto. A battery is not required because the magneto generates its own power.

Coil ignition
Low voltage current, supplied by a battery, is converted by the coil (a form of transformer) into high voltage current. The distributor supplies the sparking plugs with pulses of high voltage current.

To understand how the system works, careful reference should be made to Figure 2.12.

Low voltage current, usually 12 volts, is supplied by the battery through the ignition switch to the low tension (voltage) side of the coil. The terminal on the coil is usually marked 'sw'. The

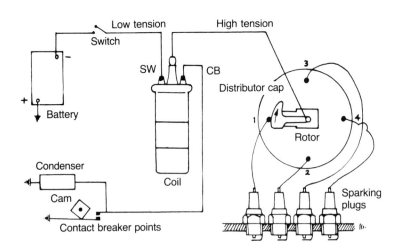

FIGURE 2.12 Diagram of coil ignition system.

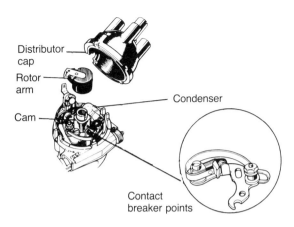

Distributor cap
Rotor arm
Cam
Condenser
Contact breaker points

FIGURE 2.13   Distributor cap and points.

current leaves the coil from the terminal marked 'cb' and flows to the contact breaker points and condenser. These are both earthed. The contact breaker points are housed in and driven by the distributor unit. A cam on the rotor arm shaft opens the contact breaker points. They are closed by a spring. When the cam opens the contact breaker points, the low tension current can no longer flow through them to earth. This causes a brief pulse of very high voltage electricity to be induced in the high tension side of the coil. This surge of current may be as high as 10,000 volts; the amperage is very low.

The condenser is a form of electrical shock absorber which reduces arcing (flashing) when the contact breaker points open. The condenser absorbs the current when the points open and uses it to boost the strength of the high voltage current induced in the coil.

The high tension current flows from the coil, along the king lead to the distributor cap. Inside the distributor cap there are four equally spaced brass terminals (for a four-cylinder engine). Each terminal is connected to a sparking plug by a plug lead.

The rotor arm is timed so that it is very close to one of the distributor cap terminals every time the contact breaker points open. The plug leads are connected to the plugs in sequence to suit the firing order of the engine. In this way, the cylinders receive a spark a few degrees before the top of the compression stroke. When the high tension current reaches the spark plug, it jumps across a gap to earth. This causes a spark which ignites the compressed mixture of fuel and air.

When an engine runs at 2,000 rpm each plug produces 1,000 sparks per minute. A four-cylinder engine running at this speed will need 4,000 sparks every minute.

Electronic ignition systems

Contact breaker points suffer from mechanical wear and pitting of the contact faces; this means regular replacement is necessary. At high speeds, the contact breaker points bounce, which results in incorrect timing of the spark. Electronic ignition systems overcome these problems. There are two main types.

The *transistor assisted coil ignition system* has breaker points but they are only used to operate a primary circuit transistor. This controls a second transistor which supplies current to the induction coil from which high voltage current is passed to the distributor and plugs. As very little current flows across the contact breaker points, this system overcomes the problem of pitting but bounce and wear still occur.

The *breakerless system* has an electronic switch and rotor unit which carries out the function of the contact breakers in a conventional system. The breakerless system therefore overcomes the problems of wear and bounce too.

The magneto
A magneto does not require a battery. It is driven by the engine timing gears. A flywheel magneto, mainly used with single-cylinder engines, is built into the engine flywheel.

Electricity is produced when an electrical conductor is rotated in a magnetic field. Alternatively, the magnet can rotate round a stationary conductor.

A flywheel magneto has a magnet which is rotated round a stationary coil (conductor); this produces the low voltage current which is provided by the battery in a coil ignition system. The low tension current is converted into a continuous supply of sparks in the same way as described for the coil ignition system.

The flywheel magneto has the contact breaker points attached to the engine housing, inside the flywheel. There is no distributor cap since the plug lead connects the high tension terminal on the coil to the spark plug.

Multi-cylinder engines with a magneto are no longer in common use on farms. In this case, the magneto is driven by the engine timing gears. It has a distributor cap to supply the current to the plugs. There is an impulse coupling in the magneto drive, which speeds up the magneto drive when starting the engine.

Sparking plugs

A sparking plug has a metal body which is screwed into the cylinder head. The centre of the plug is made of porcelain with an electrode through it, to carry the current. There is a second (earth) electrode at the bottom of the plug. The spark occurs when the current 'jumps' from the centre electrode to earth.

There are many types of sparking plug, made to suit different engine designs and working temperatures. Plugs are made with different thread lengths (reach). The engine operating temperature also decides choice of plug. Some engines need cold running plugs, others require hot running plugs. The metal plug body conducts heat away from the electrodes. A hot running plug retains more heat and, in this way, keeps the electrodes at a higher temperature. The sparking plug heat range shows whether it is a hot or cold running plug.

Always use the correct plug for the engine and make sure that a multi-cylinder engine has identical plugs in all cylinders. The wrong plug will result in poor engine performance.

Care of the ignition system

Contact breaker points The gap between the contact breaker points, when they are open, should be between 0.3 and 0.4 mm. The stationary point is adjustable. The contact breaker points must be fully open before setting the gap. The points can be removed for cleaning, but great care must be taken to keep both faces parallel. When the faces are pitted it is usually best to fit a new set of points. ·

Sparking plugs The gap between the centre and earth electrodes should be about 0.6 mm. Move the earth electrode to adjust the gap. Never attempt to move the centre electrode, because the porcelain insulator will crack.

Sparking plugs should be removed and cleaned to prevent a build-up of carbon on the electrodes.

Carbon can be removed with an emery cloth or a wire brush.

Sparking plugs have a limited working life. It is false economy to use the same set of plugs for many hundreds of hours. The gap settings given for plugs and points are only a guide. Check with the instruction book for the correct settings and renewal periods for any particular engine.

Ignition troubles

Starting or running problems are frequently caused by fuel or ignition faults. Provided that there is an ample supply of fuel and the carburettor is clean, check for the following minor ignition faults before calling in an expert:

1. When there is no spark, suspect:
   (a) Contact breaker points stuck open or shut.
   (b) Contact breaker points very dirty or pitted.
   (c) Broken, or loose king lead.
   (d) Loose terminals.
   (e) Damp distributor cap or leads.
   (f) Cracked distributor cap.
2. When the spark is very weak, look for:
   (a) Dirty spark plugs.
   (b) Dirty contact breaker points.
   (c) Loose terminals.
   (d) Dampness.
   (e) Discharged battery.

THE TWO-STROKE ENGINE

This type of engine is still in use on many farms, the most common application being to drive chain saws.

The main differences between two-stroke and four-stroke engines are:

1. Every other stroke is a power stroke.
2. The two-stroke petrol engine has no valves. Ports (openings) in the cylinder walls serve the same purpose as valves in a four-stroke engine. There are three ports:
   (a) The inlet port is connected to the fuel and air supply.
   (b) The transfer port connects the crankcase to the cylinder.
   (c) The exhaust port connects the cylinder to the exhaust pipe.

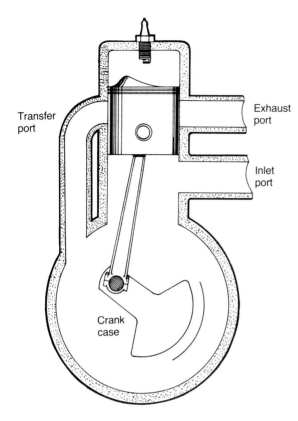

Transfer
port

Exhaust
port

Inlet
port

Crank
case

FIGURE 2.14   Section of a two-stroke engine.

3. The crankcase is airtight and does not contain any oil. Engine lubrication is by oil mist (oil is mixed with the petrol).

First stroke   As the piston nears TDC, a mixture of air and fuel (which contains some oil) is drawn into the crankcase. Some more mixture above the piston is compressed and then ignited with a spark from a flywheel magneto. Expansion from the heat produced drives the piston down. The piston covers the inlet port and the fresh mixture in the crankcase is slightly compressed.

As the piston nears BDC, the exhaust and transfer ports are uncovered. The waste gases are released to the exhaust pipe. The slightly com-

pressed mixture in the crankcase enters the cylinder through the transfer port. Some two-stroke engines have a piston with a specially shaped top which swirls the incoming mixture. This action helps to clear the exhaust gases from the cylinder.

Second stroke   The piston starts the upward stroke and the remaining exhaust gases are forced out. Next, the piston covers the exhaust and then the transfer ports. The mixture above the piston is compressed. By this time, the piston has uncovered the inlet port and a new supply of fuel and air is drawn into the crankcase. The mixture above the piston is ignited and the cycle starts again.

Two-stroke engines are air cooled. The cooling fins must be kept clean to prevent overheating. Always add oil to the petrol in the correct proportion. Check with the instruction book.

SUGGESTED STUDENT ACTIVITIES

1. Trace the path of the diesel fuel from the tank to the injectors.
2. Learn how to bleed the diesel fuel system of your tractor. Do not wait until you run out of fuel.
3. Find out the recommended change periods for the diesel fuel filters on the tractors at a farm you know.
4. Study a coil ignition system on a car engine and identify the various parts described in this chapter. Inspect the contact breaker points and sparking plugs for setting and condition.
5. Look for the two types of air cleaner and find out how to service them.

SAFETY CHECK

Use barrier cream when working with diesel fuel. This will give protection from dermatitis, an unpleasant skin inflammation which can be caused by diesel fuel.

Chapter 3

TRACTOR ENGINE LUBRICATION AND COOLING

Metal surfaces cannot be machined to a perfectly smooth state. When two dry metal surfaces run together at speed, there will be a great deal of friction resulting in rapid wear and excessive heat. By introducing a lubricant film between the two metal surfaces the wear rate will be reduced and much of the heat will be dispersed.

When an engine is running it is vital to remove much of the heat produced to ensure the moving parts do not overheat or, in extreme situations, seize. Both the lubrication and cooling systems play an important part in removing unwanted engine heat.

ENGINE LUBRICATION

The three main functions of an engine oil are:

1. To lubricate the moving parts to reduce friction.
2. To help cool the engine by removing some of the heat caused by friction.
3. To form a seal between the pistons and cylinder walls. Without this seal there would be very little compression in the combustion space.

Oil classification
Traditionally, oil has been classified according to its viscosity (thickness). Different oils are given a viscosity number; thin oils have a low number and thick oils a higher one. The viscosity numbers were devised by the American Society of Automotive Engineers (SAE). The viscosity range includes:

SAE 20 and SAE 30 – engine oil.
SAE 40 and 50 – transmission oil.
SAE 90 – for some transmission systems.

SAE 120 and SAE 140 – a very thick oil used for bearings etc.

Even the best quality lubricating oils will change their viscosity as their temperature changes. Some oils have a special additive to help keep the viscosity stable as the oil temperature changes.

Multi-grade oils are now used for most agricultural engines. They span a number of viscosity ratings, making them suitable for engines and transmission systems. They also meet the need for an oil which is thin enough for cold starting yet thick enough to be fully efficient at normal engine temperatures.

The viscosity rating only refers to the thickness of the oil and the cheapest are straight oils with no special additives. However, modern engines require a lubricant with special chemical additives to ensure a long life for both the oil

PLATE 3.1 *Checking engine oil level. The dipstick on this engine is contained in the oil filler cap.*
(Massey-Ferguson)

and the engine. A straight oil is not suitable for high performance petrol or diesel engines.

Additives
Depending on their purpose most agricultural oils contain a variety of additives. A good quality engine oil will have:

Detergent   Stops a harmful build-up of deposits which are produced when fuel is burnt causing sludge in oilways and bearings. When the oil is drained, these deposits are carried away with the oil.

Dispersant   Keeps soot suspended in the oil, avoiding deposits in the oilways, piston ring grooves, etc.

Anti-oxidant   Reduces the effects of oxidation (heat and chemical breakdown) which thickens the oil.

Anti-corrosive   Protects the engine from the corrosive action of water and acids formed when fuel is burnt.

Anti-wear   Reduces the wear rate of moving metal surfaces especially when the engine is under heavy load.

To reduce the number of grades and types of oil needed on the farm, most oil companies sell a multi-purpose oil. It has all the additives required to make it suitable for most tractor engines, transmission systems and hydraulics. Multi-purpose oil must not be used for some models of tractor as the construction or type of material used requires a special lubricant. If in doubt, always refer to the tractor instruction book.

Oil Contamination

The efficiency of lubricants, especially engine oils, is reduced with use. Diesel engines require an oil change after about 250 hours' service and petrol engines in half this period. However, the oil change period varies greatly and the instruction book should be consulted to find the correct service period for any particular engine.

The oil must be drained and renewed because it will become contaminated with:

- *Sludge*   This is produced from carbon, water and other substances formed when fuel is burnt in an engine. The sludge will block

oilways and bearings and will, in time, clog piston rings.
- *Dust* enters the engine, especially when the air cleaner and crankcase breather are not serviced regularly.
- *Metal particles*   Minute pieces of metal from the bearings and other parts of the engine will find their way into the oil, especially when the engine is new.
- *Engine heat* causes gradual breakdown of the structure of the oil. Chemical changes also occur due to the effects of oxygen (oxidation).
- *Fuel*   Small amounts of fuel will always get past the pistons and mix with the oil, especially if the engine runs below its correct working temperature.
- *Water* is produced when fuel is burnt, also from condensation, adding to oil contamination. A yellow sludge on the underside of the filler cap is an indication that the oil may be contaminated with water.

Dirty oil filters, poor storage conditions for unused oil and dirty oil handling containers will also add to the problems of oil contamination. These can be avoided by regular servicing and good storage.

Changing Engine Oil

The oil should be drained when the engine is hot. The oil filter must be replaced after thoroughly washing the filter bowl. Many engines have a screw-in-type cartridge filter which should be removed and replaced with a new one.

When the oil has been drained, replace the sump plug and fit the new filter element. Make sure that the rubber sealing ring is in good condition. If it is not, fit a new one. Refill the sump with the correct grade of clean engine oil to the full mark on the dipstick. Run the engine for a short while and then recheck the dipstick level. Finally check for leaks and record the engine hour meter reading so that you will know when the next oil change is due.

Some older engines may be lubricated with straight mineral oil. This type of lubricant has no detergent properties and in time will permit sludge to build up in the engine. After draining such engines it is not advisable to refill them with a detergent oil, as its cleansing properties will remove much of the sludge. When this happens, an engine overhaul will soon be necessary.

ENGINE LUBRICATION SYSTEMS

Force Feed

Figure 3.1 shows the layout of a typical force feed lubrication system. The oil is pumped, under pressure, around the engine.

The oil pump, usually driven by the engine camshaft or timing gears, draws oil from the sump, through a strainer, and then pumps it to the filter element. The filtered oil passes to the main bearings, camshaft bearings and the big end bearings. Oil is also pumped to the valve gear. The other parts of the engine are lubricated by the oil as it runs back to the sump. A relief valve is included in the system to protect the pump from overloading, especially at high speeds. On some engines, the relief valve ensures that oil can reach the bearings if the filter becomes blocked through lack of maintenance.

Some tractors have an oil pressure gauge. This is connected by a small pipe to the main oilway. A typical oil pressure is 2.8 kg/cm² (40 psi). All tractors have an oil warning light connected to a pressure switch in the main oilway. Oil pressure always falls as engines get hot but a very high or very low oil pressure should be investigated. Causes of low oil pressure include:

1. Low oil level.
2. Blocked oilway.

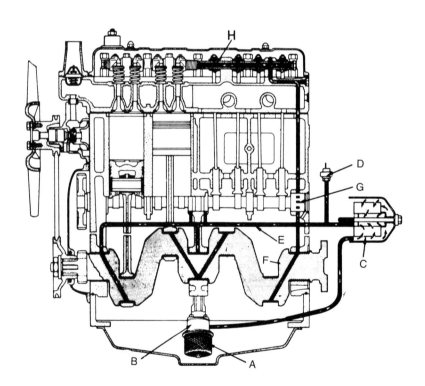

A—Inlet gauze
B—Oil pump
C—Filter
D—Warning light pressure switch
E—Oil gallery
F—Crankshaft oil passage
G—Camshaft bearing
H—Valve rocker shaft

FIGURE 3.1   Engine lubrication system.

3. The oil is too thin. This may be because:
   (a) The engine is overheating.
   (b) The oil is diluted with fuel.
   (c) The wrong grade of oil is being used.
4. Worn oil pump.
5. Worn engine bearings.
6. Relief valve weak, or stuck open.
7. Blocked oil filter. Only if the pressure gauge is after the filter in the flow circuit.

Causes of high oil pressure include:

1. Blocked oilway.
2. Relief valve stuck in the closed position.
3. Oil too thick. This may be because:
   (a) The engine is running too cold.
   (b) The oil is very dirty.
   (c) The wrong grade of oil is in the engine.
4. Blocked oil filter (when the gauge is in front of the filter in the oil circuit).

Crankcase breather
This is a small air vent, with a filter, which prevents a build-up of pressure in the sump. It is often built into the oil filler cap. Pressure is more likely to build up in old engines, where gases can pass between the pistons and the cylinder walls. The gases must be released from the sump as they contain small quantities of water.

The crankcase breather must not be allowed to become blocked. The filter in the breather cap should be washed in petrol.

Splash Feed Lubrication

Used for small single-cylinder engines. The moving parts are lubricated by the oil as it is splashed over them, by a small scoop attached to the big end bearing cap. Some engines have a simple impeller which serves the same purpose. The oil in the crankcase must be kept at the correct level as there is not a pump to ensure proper lubrication.

Oil Mist Lubrication

This is used for two-stroke petrol engines. Because the crankcase must be dry and airtight, the engine is lubricated with oil mixed in the petrol. This creates an oil mist which lubricates the moving parts. A typical engine requires a mixture of 20 parts of petrol to 1 part of oil. Check with the instruction book to find the correct mix for any engine you may use.

The crankcase of a two-stroke engine sometimes has a drain plug. This is used to remove any oil which may collect. Some engines are a problem to start as there is too much liquid in the crankcase.

Too much oil mixed with the petrol will cause excessive carbon deposits in the engine and a lot of blue smoke will appear from the exhaust.

ENGINE COOLING

Engines create a vast amount of heat. This must be removed if the engine is to continue operating efficiently.

Thermo-syphon Cooling

This is only used for water-cooled stationary engines. It depends on convection currents to keep the cooling water circulating. Hot water, from around the cylinder, rises to the top of an open header tank. The surrounding air cools the water which is then replaced by more hot water rising from the area of the cylinder. In this way, the water keeps circulating to cool the engine.

Impeller-assisted Cooling

The principle of the thermo-syphon system is still used, but an impeller pump ensures that the water circulates properly. Figure 3.2 shows the arrangement of the water-cooling system. Water rises up through the cylinder block as it becomes hot and then flows down through the radiator as it cools. A fan draws cold air through the radiator and the impeller assists with the circulation of water.

The thermostat controls the water temperature. When the engine is cold, the thermostat prevents circulation through the radiator. The water can only rise up through the engine to the impeller, which returns the water to the cylinder block through the by-pass hose. When the engine cooling water reaches a temperature of about 75°C, the thermostat valve opens and water can flow into the radiator for cooling. The thermostat

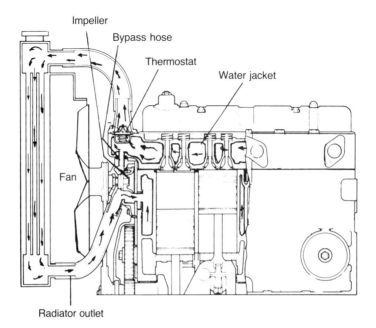

FIGURE 3.2   Engine cooling system.

ensures that the engine warms up quickly from a cold start.

The radiator is pressurised. This raises the boiling point of the cooling water. When removing the radiator cap from a hot engine, turn the cap very slowly to avoid the risk of scalding. It is best to cover the cap with some sacking before undoing it.

Most cooling systems have two drain taps, one on the cylinder block and the other at the bottom of the radiator.

Some tractors have a small expansion tank connected by a small rubber tube to the radiator. This is used to maintain the water level in the radiator. When topping up the cooling system, the additional water is poured into the expansion tank. All pressurised radiators have an overflow pipe if they are not equipped with an expansion tank. When the radiator pressure is higher than normal, a relief valve, built into the radiator cap, opens and releases the pressure (sometimes water as well) through the overflow pipe.

A typical cooling system will have a radiator cap with a valve which opens at a pressure of 0.5 kg/cm^2 (7 psi).

Frost protection
An anti-freeze solution, ethylene glycol, is used to protect the cooling system from frost damage. A 25 per cent solution of anti-freeze and water will protect the system down to a temperature of about −18°C. A 25 per cent solution consists of 2 litres of anti-freeze and 6 litres of water.

Anti-freeze does not prevent freezing, it only lowers the freezing point of the coolant. It lowers the boiling point too. It is best to top up the cooling system with a ready mixed anti-freeze solution during the winter months.

Engine-operating temperature
A cold, or overheated engine, cannot work properly. The cooling water should be at a temperature of approximately 90°C to ensure efficient operation. Keep an eye on the temperature gauge on the instrument panel.

Care of the Cooling System

- Check radiator water level, daily in hot weather, otherwise weekly. Top up when necessary. The water level should be about

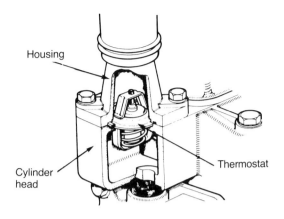

FIGURE 3.3 Thermostat.

PLATE 3.2 *The cooling fins of an air-cooled engine.*
(Ransomes, Sims and Jefferies)

25 mm below the neck to allow for expansion.

Never add large quantities of cold water to a very hot engine. The cast iron cylinder head or cylinder block will crack.

- Flush the cooling system to clear any sediment. This can be done when the cooling system is drained, ready for a new solution of anti-freeze.
- Keep the outside of the radiator clear of dirt and chaff. Remember to wash or blow the dirt out the way it came in, by using a hose pipe

or airline; work from the fan side of the radiator.

- Maintain the correct fan belt tension. Adjustment is explained in the section dealing with the charging system.
- Check for leaks, make sure hose connections are tight and watch for perished hoses.
- Very few water pumps need any lubrication. Those which do should be greased weekly.

Causes of engine overheating
1. Low water level.
2. Radiator blocked. This can be either inside or outside of the radiator.
3. Fan belt slack.
4. Thermostat stuck shut.

Air Cooling

Some tractors and almost all small engines are air cooled. There is no chance of frost damage but air-cooled engines will overheat if the cooling fins are not kept clean.

Fan blades on the engine flywheel blow air over the fins on the cylinder head and block. The fins give a much larger surface in contact with the air to help disperse the engine heat. Air-cooled tractor engines have a separate blower to create a flow of air over the cooling fins.

Transfer of Heat

Heat can be transferred from one point to another in three ways:

Conduction Hold a piece of metal in the fire and, before long, the end in the hand becomes warm. Heat moving through metal in this way is transferred by conduction. Metals are good conductors of heat, especially copper and brass.

Convection Hot air will rise above cold air because it is less dense (lighter). Hot water rises too. Heat is transferred by convection when hot water or air rises above cold water or air. A domestic convector heater works on this principle.

Radiation Rays of heat can be felt as they leave a fire. These rays always travel in straight lines from the heat source. This form of heat transfer is called radiation.

When an engine is cooled, the heat is conducted through the cylinder walls into the cooling water. Convection occurs as the hot water rises from the cylinder block to the radiator where the water is partly cooled by radiation. Heat is also radiated from the exhaust manifold.

SUGGESTED STUDENT ACTIVITIES

1. Locate the engine oil filler, the dipstick and the oil filter element.
2. Check the instruction book to find how often the oil and the filter element should be changed on various tractors.
3. Read the labels on oil drums to find as much information as you can about the contents.

What is the viscosity and has the oil any additives?
4. Read any information you can find about oil refineries and how different types of fuel and lubricant are made.
5. Check the instruction book to find the capacity of the cooling system.

SAFETY CHECK

Always take care when removing the radiator cap from a hot engine. It should be removed slowly, preferably with a piece of thick material placed over the cap, to protect yourself from very hot or scalding water.

Chapter 4

TRACTOR ELECTRICAL SYSTEMS

The tractor battery must supply the electricity required to operate the numerous items of electrical equipment found on tractors and self-propelled machines. It is also required to supply power for the remote controls on a variety of implements. This chapter deals with charging and starting systems.

THE BATTERY

Batteries store direct current (DC). Mains electricity is alternating current (AC); this form of electricity cannot be stored in a battery unless it is rectified (changed) to DC first. This is the function of a mains battery charger. It rectifies AC supply to DC and also transforms mains voltage to battery voltage, normally 12 volts.

Electricity generated by the alternator, or on some tractors by the dynamo, is stored in the battery. It consists of a hard rubber or polypropylene box containing a number of lead plates immersed in sulphuric acid. Electrical energy from the charging system is converted into chemical energy by the battery and stored in that form. During the charging process, both hydrogen and oxygen are given off and the electrolyte (sulphuric acid) becomes stronger. The acid strength weakens when the battery discharges.

When the driver uses any of the tractor's electrical equipment, the chemical energy is changed back to electricity. When the engine is running, the charging system provides most of the electric power needed to operate the lights, horn, screen wiper, radio, etc.

Battery maintenance
Most tractors have a battery with a screw cap to each cell. This type requires regular mainten-

ance, as described below. Many polypropylene cased batteries are maintenance free or have ultra-low maintenance requirements. The casing is translucent and the electrolyte level can be seen. If it is below the minimum level mark the caps can be pierced with a screwdriver and then unscrewed. It is then possible to top up the electrolyte in the normal way.

Many batteries are ruined through neglect. A little attention will ensure long, trouble-free service. For maximum battery life:

1. Remove the cell caps and check the electrolyte level: it should be just above the plates. If it is low, use distilled water to restore the electrolyte to the correct level. Tap water

PLATE 4.1 *Withdrawing the battery from its housing. Note the cover over the terminal to reduce corrosion.* (Massey-Ferguson)

should not be used. Remember to clean the top of the battery before removing the cell caps.

2. Keep the vent holes in the cell caps clear.
3. Make sure that the terminal connections are clean and tight. A light coating of petroleum jelly will reduce corrosion.
4. The earth connection to the tractor chassis must be tight. The negative terminal is connected to earth. Some tractors, mostly those made before 1965, have a positive earth connection.
5. Keep the top of the battery clean and dry.
6. The battery must be secure in its mounting, to prevent damage to the casing.
7. Remove the battery from a tractor or machine which will not be in use for a long period. Store the battery in a cool shed and keep it well charged. A discharged battery will freeze.

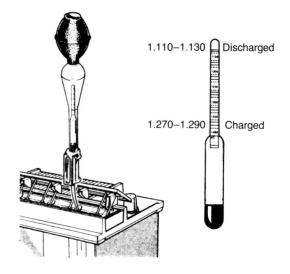

1.110–1.130 Discharged

1.270–1.290 Charged

FIGURE 4.1   Testing a battery with a hydrometer.

Battery testing

A hydrometer is used to check the state of charge of a battery. It measures the specific gravity (concentration) of the acid. A float in the hydrometer body has either three coloured bands or a scale to indicate the strength of the electrolyte. The coloured bands show full charge, half charge or discharged.

The specific gravity scale readings are:

Specific gravity	State of charge
1.280	Full charge
1.200	Half charged
1.110	Discharged

To use a hydrometer, remove the cell cap and put the rubber tube into the electrolyte. Squeeze the rubber bulb and then gently release it to draw some electrolyte into the hydrometer body. Note the specific gravity reading, or state of charge, and return the liquid to the cell. Repeat this procedure for at least two more cells. The electrolyte is acid so keep it away from your clothes and skin. After use, rinse the hydrometer in clean water.

Battery charging

A discharged battery can be recharged with a mains battery charger. This converts the mains supply from AC to DC. The charger may have an adjustment which offers the choice of 12, 6 or 2 volts; others have only a 12 volt output.

Use the following procedure to put a battery on charge:

1. Make sure the battery top is clean, remove caps and top up the electrolyte if necessary. Leave the cell caps slack, to prevent gas pressure building up in the battery.
2. Connect the charger clips to the battery terminals, making sure the positive and negative clips are on the correct terminals. Check that the charger is set to suit the battery voltage.
3. Plug the charger into the mains supply and switch on.
4. Check the electrolyte every few hours. When fully charged, switch off the charger at the mains supply before removing the terminal clips. Replace and tighten the cell caps.
5. Because a battery gives off hydrogen and oxygen when under charge, avoid the use of naked flames. *Never use a lighted match to check the battery.*

Using jump leads

Tractor batteries are sometimes found to be discharged when the starter switch is used. Jump leads can be used to overcome this problem in an emergency, but if the battery is faulty it must be replaced.

Care must be taken when using jump leads to prevent damage to the electrical system and personal injury. For negative earth systems, use the following sequence:

1. Connect the positive terminals of both batteries with one jump lead.
2. Connect the other lead to the negative terminal on the booster battery.
3. Using a good earth point on the tractor, which is at least 500 mm from the battery, connect the opposite end of the negative lead.
4. Ensure the leads are clear of any moving parts and then operate the starter. Once started, remove the leads in the reverse order.

It is a dangerous practice to connect the jump lead to the negative terminal of the discharged battery. This action can result in severe arcing which can have explosive results. Never allow the jump lead clips to touch together once they have been connected to the booster battery.

THE CHARGING SYSTEM

An alternator or a dynamo is used to generate electricity for the battery. The dynamo has been replaced by an alternator on current models of tractor. However, many farm tractors are still in use which rely on a dynamo to generate electricity. The advantages of the alternator are that it has a high output at low engine speed and it has a higher maximum output than a dynamo of equivalent size.

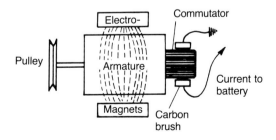

FIGURE 4.2 Principles of a dynamo.

When an electric conductor moves in a magnetic field, current is created. The dynamo has stationary magnets and a rotating conductor (armature). The alternator differs in that it has static conductor windings (stator) and a rotating magnet (rotor).

The Dynamo

The armature consists of a large number of conductors, each insulated from the next. It rotates in a magnetic field created by two electro magnets. Current is supplied by the battery

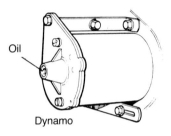

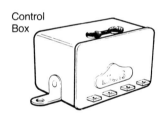

FIGURE 4.3 Dynamo and control box.

to energise the magnets. This is called the field current. When the armature is rotated—driven by the fan belt from the engine crankshaft pulley—the current generated passes to the commutator. Here it is collected by a carbon brush, held against the commutator by spring pressure, which feeds it to the battery. There are two carbon brushes; one is earthed and the other is connected to the dynamo output terminal.

It is necessary to control dynamo output and prevent the battery discharging when the engine is stopped. A control unit is built into the charging system to achieve this. The voltage regulator prevents overcharging. It regulates the strength of the field current which creates

the magnetic field. When the battery is not fully
charged the regulator provides a strong magnetic
field, resulting in a high rate of charge from
the dynamo. When the battery is fully charged
the field current is reduced, thereby cutting the
charging rate to a minimum.

When the engine stops, current must be prevented from flowing back to the dynamo because,
if this happens, the field windings will be damaged.
The cut-out, also to be found in the control unit,
is an automatic switch which prevents back flow to
the dynamo.

The Alternator

The working principle is the same as that for a
dynamo but the conductor is stationary and the
magnets rotate.

Current is supplied to the rotor windings by
two carbon brushes to create the magnetic field.
Alternating current induced (created) in the
stator windings flows to a rectifier in the alternator housing where it is changed to direct current
and passed to the battery. A voltage regulating
mechanism is built into the circuit. The cut-out

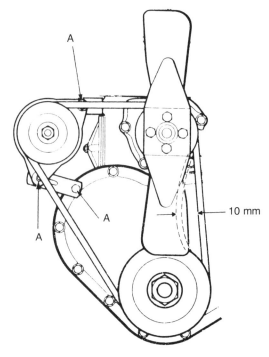

A – Adjusting bolts

FIGURE 4.4   Fan belt adjustment.

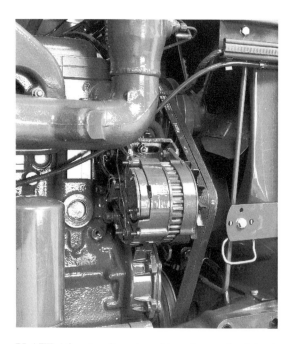

PLATE 4.2   *An alternator, driven by two fan belts. It
is pivoted outwards to increase belt tension. The
cooling fan can be seen in the background.*

(Massey-Ferguson)

is replaced by a relay switch which prevents back
flow from the battery to the alternator.

Simple charging faults
The ignition warning light warns the driver
that the generator is not charging the battery. If
the warning light comes on when the engine is
running normally, possible causes include:

1. Loose or broken wires or connections.
2. Fan belt slack.
3. Worn or dirty carbon brushes.

Maintenance of the charging system
Correct tensioning of the drive belt is very
important. The generator is normally driven by
the fan belt, but on some engines the generator
has a separate drive belt. A slack fan belt can
cause overheating of the engine as well as a
low rate of charge. An over-tightened belt will
damage the generator bearings.

To adjust the fan belt, slacken the securing

PRE-ENGAGED STARTER

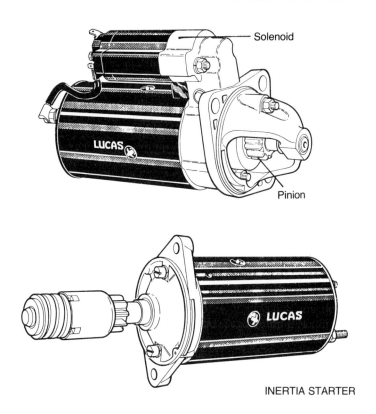

FIGURE 4.5   Inertia and pre-engaged starter motors.

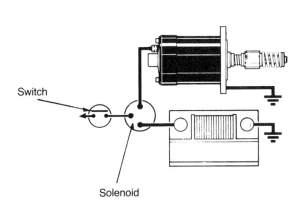

FIGURE 4.6   Starter circuit.

bolts and move the generator on its pivots until the correct tension is obtained. Finally retighten the bolts and check tension.

Generators have an internal fan to keep them cool, but the air vents must be kept clean to allow maximum airflow. Most dynamos need occasional lubrication with a drop of light oil in the bearing end cap. No other attention is needed apart from ensuring the terminal connections are clean and tight.

STARTING SYSTEMS

There are two main types of starter motor used on tractor engines. The inertia type has the starter pinion (gear) turning at full speed before it engages with a ring gear on the engine

PLATE 4.3   *Four-wheel-drive tractor equipped with a full set of road and working lights. An amber flashing lamp is attached to the cab roof.*   (Massey-Ferguson)

flywheel. The pre-engaged starter motor has its pinion engaged with the flywheel ring gear before it begins to turn.

Many multi-cylinder spark ignition engines have an inertia-type starter motor. When the motor turns, the starter pinion moves along its shaft on a helix (very coarse thread). The pinion engages with the ring gear and turns the engine. When the engine starts, the pinion is thrown out of mesh from the flywheel ring gear. At this point, the starter switch is released and the pinion returns to the rest position as the motor slows down.

Diesel engines need the pre-engaged-type starter. The pinion is fully engaged with the fly-

wheel ring gear before the motor is switched on. This method of engagement reduces the wear rate of the pinion and ring gear. The pre-engaged starter is a heavy duty unit of reasonable size which has the necessary power to turn a diesel engine at a cranking speed of around 200 rpm. All modern pre-engaged starters have a solenoid to engage the pinion with the flywheel ring gear and then complete the circuit from the battery to the motor.

Some early models of diesel-engined tractor have a hand lever to engage the pinion with the flywheel ring gear. Further movement of the hand lever energises a solenoid which completes the starter circuit.

A solenoid consists of a coil in a housing with a soft iron plunger through the centre. When the coil is energised with current from the battery, the magnetic field created draws the plunger along the coil. There are two terminals on a starter solenoid casing: one is connected to the battery and the other to the starter by means of heavy cable. Movement of the plunger engages the starter pinion with the flywheel ring gear. At the end of its travel, the solenoid plunger connects the two terminals, allowing current to flow to the starter motor.

When the engine starts and the switch is released, the solenoid is no longer energised so it returns to rest. In so doing, it disconnects the electrical supply and withdraws the pinion from the flywheel with the help of a spring.

The only driver maintenance required is to keep the battery, solenoid and starter terminals clean and tight.

Lights and Fuses

All tractors have fuses to protect the electrical circuits from overload. The fine wire in the fuse cartridge will melt (blow) if too much current flows through the circuit. This prevents damage to the various items of electrical equipment such as lights, indicators, horn, radio, etc.

The flow of electric current in a circuit is measured in amps. A fuse is rated according to how many amps the fuse wire will carry before it blows. Try to find why a fuse has blown before replacing it and always use a fuse of the same rating as the original. A bigger fuse will not give the proper protection to the equipment and for this reason should not be used.

Tractors are equipped with a full set of lights which comply with the Road Traffic Regulations. Additional lights are usually fitted to aid working in the field after dark. Indicators and hazard warning lights are also standard equipment. All lights should be checked frequently and faulty bulbs replaced. It is a legal requirement that where lights are fitted, they are in working order at all times.

Suggested Student Activities

1. Remember to service your tractor battery at regular intervals. Check the electrolyte level and keep the terminal connections clean and tight.
2. Learn how to use a battery testing hydrometer and interpret the float readings.
3. Find out how to adjust fan belt tension. Check the instruction book for the correct setting.
4. Locate the starter, solenoid, alternator and fuse box on your tractor.
5. Learn and observe the correct procedure when using jump leads to start an engine.

Safety Check

Never use a naked flame, or smoke, near a lead acid battery when it is being charged. An explosive mixture of hydrogen and oxygen are given off during the charging process.

Chapter 5

TRACTOR TRANSMISSION SYSTEMS

Tractor engines run at speeds in excess of 2,000 rpm. In the lowest gear, the rear wheels will turn at approximately 20 rpm. The purpose of the transmission system is to reduce the engine crankshaft speed by means of the gearbox, crown wheel and pinion and the final reduction gears to give a forward speed suitable for the job in hand.

A tractor in low gear can pull a heavy load. The engine would stall if the driver attempted to pull the same load in a high gear. The gearbox

PLATE 5.1 *Drive shaft to front wheels.* (Massey-Ferguson)

44

provides the driver with a wide option of forward speeds to suit all types of activity from heavy field work to high speed transporting.

In simple terms, the power available is equal to the load multiplied by the forward speed. So, in field conditions, the same engine power can be used to pull heavy loads at low speed and light loads at high speed. However, the same tractor can pull heavier loads on the road in high gear because the surface offers very little resistance compared with a wet field.

Four-wheel-drive increases the pulling power of farm tractors. Most tractors with four-wheel-drive have front wheels smaller than those at the rear. Equal size four-wheel-drive models are also made, especially at the higher end of the power range.

Line of drive
The engine flywheel is connected to the gearbox by the clutch, which is used to engage and disengage the drive to the transmission. The gear levers are used to select the forward and reverse gears. Power from the gearbox is put through a right angle drive by the crown wheel and pinion which also carries the differential. This allows the rear wheels to turn at different speeds when the tractor turns a corner. Power from the differential is transmitted by the half-shafts to the rear wheels. Many tractors have a final reduction unit on the half-shafts, which gives a further speed reduction before the power reaches the wheels.

The power take-off and the hydraulic pump are driven by the gearbox. Some tractors have the hydraulic pump mounted on, and driven by, the engine. This makes the hydraulic system completely independent of the transmission.

THE CLUTCH

As well as engaging and disengaging the drive, the clutch enables the driver to move off smoothly. The main clutch assembly is bolted to the engine

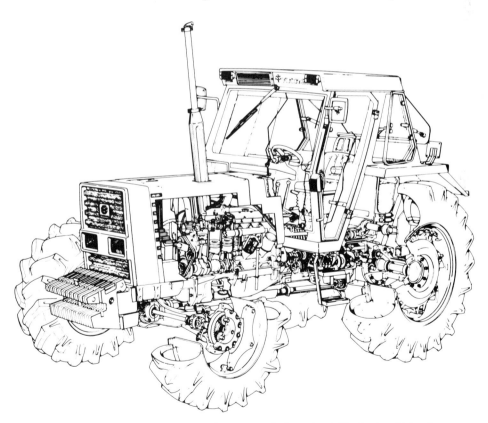

FIGURE 5.1   Cutaway view of the engine and transmission system of a four-wheel-drive tractor (*Fiatagri*)

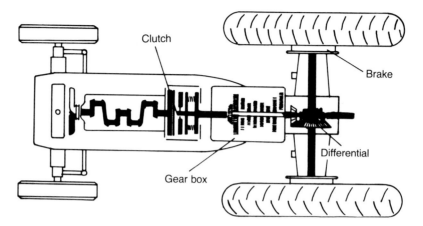

FIGURE 5.2   Layout of simple transmission system.

flywheel. With the exception of the clutch plate all the clutch components must turn with the flywheel. The clutch plate is carried on a splined shaft connected to the gearbox. It has a friction disc on both faces. The flywheel and pressure plate have smooth steel faces.

When the clutch pedal is up, the clutch disc is held against the flywheel by the pressure plate and clutch springs. The spring pressure is taken off the clutch disc when the clutch pedal is pushed down. This breaks the drive to the gearbox. The pressure plate, springs and clutch cover continue to turn with the flywheel but the clutch disc is stationary.

The clutch pedal linkage operates a thrust bearing fitted on the shaft from the clutch disc to the gearbox. When the pedal is pushed down, the thrust bearing pushes the release levers towards the flywheel. The levers compress the clutch springs and the clutch plate is no longer gripped between the flywheel and the pressure plate.

Almost without exception, tractors have a live power take-off. This means that the drive to the power shaft is not broken when the clutch disengages the drive to the rear wheels. There are two ways of achieving live power take-off drive:

Independent clutch   This arrangement has a single stage transmission clutch on the engine flywheel. The power take-off is controlled by a separate multi-plate clutch in the transmission housing. It is usually hydraulically operated and is controlled by a hand lever.

Dual clutch   Although common a few years ago, the dual clutch has been replaced by the independent clutch system of power take-off control on the latest models of tractor. The dual clutch works in the same way as a single-stage clutch. However, there are two clutch discs, two pressure plates and two sets of springs. When the pedal is pushed half-way down, spring pressure is released from the first clutch disc and the drive to the wheels is broken. The power shaft still turns. The pedal is pushed fully down to disengage the second stage of the clutch and stop the power take-off.

Clutch Pedal Free Play

The clutch pedal must have free movement before any spring resistance is felt. The pedal free play ensures complete clutch engagement. The pedal linkage can be adjusted to obtain the correct amount of free movement.

The distance varies with different models of tractor. A typical setting is 18 mm. Check for free play by pushing the pedal down with the hand.

Too little free play may cause:

1. Wear on the clutch thrust bearing.
2. Clutch slip. This is because the pressure plate does not grip the clutch plate firmly against the flywheel.

Too much free play will make gear engage-

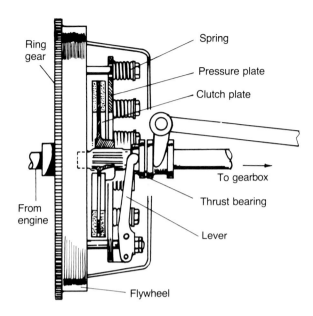

Ring gear

Spring

Pressure plate

Clutch plate

To gearbox

From engine

Thrust bearing

Lever

Flywheel

FIGURE 5.3   Single plate clutch.

ment difficult because the thrust bearing cannot completely break the drive from the flywheel.

Never use the clutch pedal as a foot rest, because the weight of your foot will take up the free play and there will be unnecessary wear of the thrust bearing.

The recent introduction of quiet cabs has resulted in the use of hydraulically operated clutches similar to those used for motor cars. The clutch pedal operates a master cylinder (a small pump and fluid reservoir) which supplies fluid, under pressure, to a ram cylinder on the clutch linkage.

The ram moves the clutch levers and the spring pressure is taken off the clutch plate to disengage the drive. When the pedal is released, the clutch springs re-engage the drive. At the same time, the fluid is returned to the master cylinder.

THE GEARBOX

The gearbox provides a range of forward and reverse speeds, as well as a neutral position. The driver usually has a choice of speeds ranging from less than 1.6 km/h (1 mph) to about 32 km/h (20 mph). Choice of gear will also determine how much the tractor can pull.

A simple gearbox has six or eight forward speeds and two reverse. Other gearboxes have as many as sixteen forward speeds and eight, or even more, reverse gears. Selection is by means of two gear levers. The first is used to select the required forward or reverse gear; the second selects the speed range. Some gearboxes offer the choice of a high or low ratio (speed range). Others have high, medium and low ratio positions for the speed range lever. Remember that one gear lever must be in neutral before most models of tractor can be started.

The range of gears available with two levers can be doubled in gearboxes which have a hydraulic or electrically operated reduction unit. A lever, switch or pedal is moved from high or low while the tractor is still moving. This gives an automatic change of speed without the use of the clutch.

Synchromesh gearboxes are now fitted to many models of tractor. This type of gearbox has been used on motor cars for many years. It enables the driver to change gear on the move. The synchromesh gearbox has special couplings which bring pairs of gears, about to be meshed together, to the same speed before the teeth are engaged.

Tractor gearboxes may have synchromesh on all forward gears. Sometimes synchromesh units are only used for the highest gear in each ratio.

A change-on-the-move gearbox can be fitted to tractors. It is a form of semi-automatic gearbox. There are two parts; a range gear lever is used to select two, or three, forward and one reverse range of speeds. The speed range is selected by using the clutch and then engaging the required ratio or reverse. The semi-automatic part of the gearbox, usually with four speeds, can be changed on the move without the use of the clutch. When the driver moves the gear lever, the gear change is carried out hydraulically.

Hydrostatic gearboxes give an infinite range of speeds from a very slow creep to a high road speed. There is a range lever to select high, low or reverse. The hydrostatic lever is moved in a quadrant to set the forward speed. As the lever is moved towards the maximum setting, the forward speed is gradually increased. Reverse speed is increased in the same way. The hydrostatic drive consists of an engine-driven oil pump which supplies oil to an hydraulic motor in the

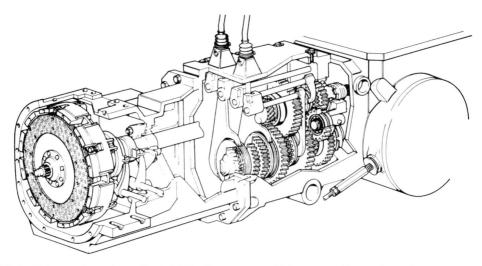

FIGURE 5.4   Sliding mesh gearbox, with straight tooth spur gears which are moved into and out of mesh with the two gear levers. The gear selector forks can be seen.   (*Fiatagri*)

transmission housing. The gear lever controls the flow of oil to the motor. Tractor speed increases as oil flow increases.

Hydrostatic drive tractors do not have a clutch as there are no gears to change. A pedal is provided to control movement in very confined spaces or when hitching implements. This acts in a similar way to a clutch but is only used on tight corners.

Sliding mesh gearboxes have straight-toothed (spur) gears. When the driver changes gear, selector forks operated by the gear lever move certain gears along on their shafts. Spur gears are rather noisy in operation.

Constant mesh gearboxes may have either spur or helical gears. They are always in mesh and are carried in groups on the gearbox shafts. The groups of gears can be connected in different sequences between the top and bottom shaft with special couplings operated by the gear lever.

Helical gears are quieter running than spur gears. They also give a stronger drive but are more expensive to manufacture.

Gearboxes are usually lubricated with an SAE 40 or 50 gear oil. The oil level should be checked every week. Some tractors have a combined gearbox and transmission housing with one supply of oil to lubricate the entire system.

THE CROWN WHEEL AND PINION

The drive from the gearbox must be put through a right angle because the wheels run at 90 degrees to the engine crankshaft. The crown wheel and pinion change the direction of the drive. The small bevel pinion on the output shaft from the gearbox drives the much larger crown wheel. When a small gear drives a larger one, there is always a reduction in speed. This means that the crown wheel and pinion turns the drive through a right angle and gives a considerable reduction in speed.

The crown wheel and pinion bevel gears may have either straight or helical teeth. Most tractors have the stronger and smoother running helical toothed bevel gears.

THE DIFFERENTIAL

If the axle was one solid shaft with the wheels secured to it, it would be impossible to turn a corner with a tractor. The differential allows one wheel to turn faster than the other when cornering. On a straight course, the differential passes equal power to both wheels.

Figure 5.7 shows a differential unit fixed to the crown wheel. A bevel gear on each

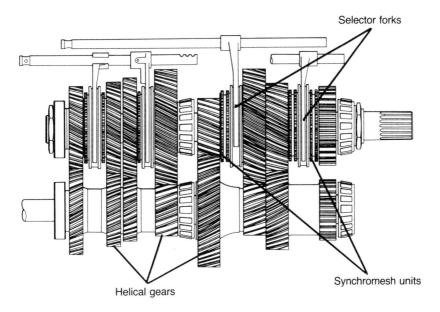

FIGURE 5.5   A constant mesh gearbox system using helical gears. Different speeds are selected by connecting various groups of gears. The gear train illustrated has synchromesh units.   (*Fiatagri*)

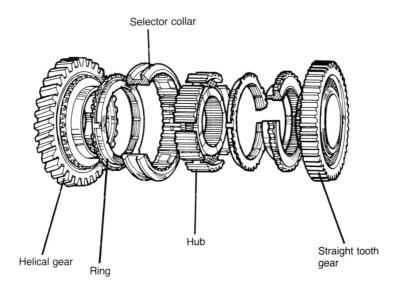

FIGURE 5.6   Synchromesh unit: the two gears are on separate shafts which are brought to the same speed by the synchromesh unit before the drives are connected. The centre hub brings both rings to the same speed when moved by the selector collar. When the shaft speeds are the same, the drive is connected.   (*Fiatagri*)

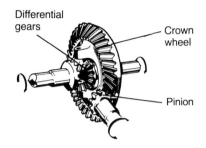

FIGURE 5.7   Principles of the differential.

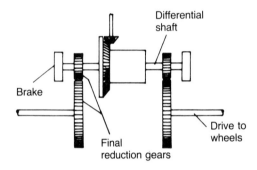

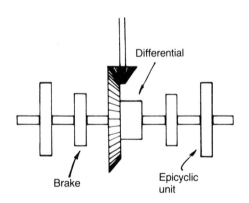

FIGURE 5.8   Alternative final drive layouts.

half-shaft meshes with the smaller differential pinions running in journals attached to the crown wheel.

When the tractor travels in a straight line, the crown wheel, half-shaft and differential gears all turn as one assembly. When the tractor turns a corner, the inner wheel slows down. As a result, the differential gears move around the slower half-shaft gear. This action increases the speed of the other half-shaft gear which is connected to the outer wheel. Because the half-shaft pinion turns faster, the tractor wheel must also turn faster as the tractor turns the corner. As soon as the tractor is back on a straight course, the half-shaft and differential gears stop turning and both half-shafts run at the same speed.

When one rear wheel spins at high speed and the other is stationary, the differential is in operation. One half-shaft gear is running at twice the normal speed and the other does not move at all.

The Differential Lock

Known as the diff-lock, this device cuts out the action of the differential when the pedal is pushed down. The two half-shafts are locked into a solid shaft and the differential and half-shaft gears cannot turn. This makes it impossible for one wheel to spin when the tractor is working in difficult conditions.

Remember that it is not possible to drive a tractor round a corner when the diff-lock is engaged.

FINAL REDUCTION GEARS

The transmission system shown in Figure 5.8 does not have final reduction gears. The rear wheels are directly connected to the differential. This type of transmission has a high reduction at the crown wheel and pinion.

Many tractors have some form of speed reduction between the crown wheel and pinion and the rear wheels. One system has a pair of reduction gears inside the transmission housing. A small gear on the shafts from the differential drives much larger gears on the shafts connected to the wheels.

Another system has the reduction gears in housings on the rear axle. Some tractors have a pair of gears in these axle housings. A small gear drives a much larger gear in each reduction unit. This reduces the half-shaft speed before the drive reaches the rear wheels.

Other tractors have an epicyclic reduction unit in each axle housing. These also reduce the speed of the half-shafts before the power arrives at the rear wheels.

Remember that as well as reducing shaft

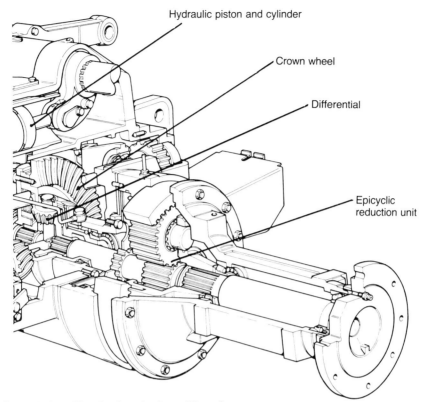

FIGURE 5.9 Transmission with epicyclic reduction. (*Fiatagri*)

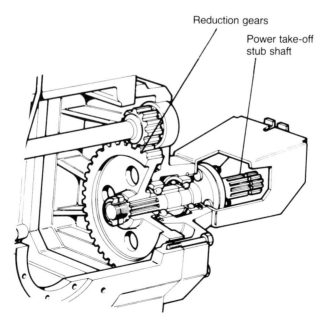

FIGURE 5.10 The power take-off stub shown is for 540 rpm operation.
It is removed and replaced with a larger stub shaft when using 1,000 rpm.
(*Fiatagri*)

speed, the reduction units give an increase in power. The power, or turning force, in the half-shafts is called torque. There are a number of places in the transmission system where the shaft speed is decreased and the torque is increased. These include the gearbox, the crown wheel and pinion and the final reduction unit.

POWER TAKE-OFF

The British Standard power take-off speed is 540 rpm. The shaft has 6 splines and has a diameter of 35 mm. There is a second (American) standard speed of 1,000 rpm. This is used to drive field machines with a very high power requirement. The shaft has 21 splines, and a diameter of 44 mm. Tractors equipped with a two-speed shaft have a high/low ratio gearbox with a lever to select 540 or 1,000 rpm.

Ground speed power take-off has a shaft speed related to the forward speed of the tractor. When the tractor is reversed with the ground speed power take-off engaged, the shaft will also run in reverse. Ground speed power take-off is used where the speed of the drive shaft to the machine must increase or decrease at a controlled rate to match changes in tractor forward speed. A trailer with power-driven axle is an example of its use.

The power take-off shaft rotates in a clockwise direction when viewed from the back of the tractor. Drive is engaged with a lever. The clutch pedal must be pushed down before engaging the drive, if the tractor has a dual clutch. There is no need to use the clutch pedal when the power take-off is controlled by an independent clutch.

There are still some tractors in use which have a single stage clutch. One movement of the clutch pedal will stop and start both forward travel and power shaft.

It is important to ensure that the inner part of the power drive shaft has a minimum overlap with the outer part of at least 150 mm when the shaft is fully extended, i.e. with the implement at its lowest position. Check also that the outer shaft is slightly shorter than the inner. This will avoid damage to the shaft when the implement is fully raised.

There are a number of legal requirements concerning the guards for power take-off shafts.

A shield must cover the top and the sides of the splined shaft; it must be capable of supporting a weight of at least 114 kg. When the power take-off shaft is not in use, this shield may be removed but must be replaced with a dust cover of equal strength.

The drive shaft from the tractor to the implement must be covered on all sides up to the first fixed bearing on the machine. Make sure the guards are kept in good condition.

BELT PULLEYS

There is little work for a tractor belt pulley in modern farming. Electricity and small stationary engines provide the power for most fixed equipment in farm buildings. A belt pulley unit can be fitted to the power take-off shaft of some tractors. A few aged tractors have a belt pulley attached to and driven by the gearbox.

There is a British Standard belt speed of 945 m/min (3,100 ft/min). Tractors which can be equipped with a belt pulley have a mark on the proof meter to indicate the engine speed which gives the standard belt speed. Most pulleys have a convex (outward curved) face to help keep the belt running centrally on the pulley. A lever on the gearbox engages the drive to the side-mounted pulley. The power take-off lever engages a rear-mounted pulley drive.

Safety regulations also apply to the use of a belt pulley. The belt run-on point must be guarded. It must also be possible to disengage the drive to the machine from the working position. Before you use any belt-driven equipment, make sure you know the requirements of the Farm Safety Regulations for belt pulleys.

BRAKES

Tractors may have expanding shoe or disc brakes. A brake is always more effective when fitted on a fast-moving shaft with a low torque. Tractors with a final reduction unit have the brakes fitted on the faster shaft between the differential and the reduction gears (see Figure 5.9). Other transmission layouts have the brakes on the rear wheel hubs.

PLATE 5.2   *The two sizes of power take-off shaft. The smaller 540 rpm shaft is in position and the larger 1,000 rpm shaft is shown for comparison. A retaining clip must be removed before the shaft can be withdrawn and replaced with the other one. It is then secured by replacing the retaining clip.*   (Massey-Ferguson)

Expanding Shoe Brakes

A pair of brake shoes, both fitted with a friction lining, are attached to a back plate. The shoes are pulled together with springs. There is a cam, operated by the brake pedal, between one end of each shoe.

The brake drum, secured to the driving shaft, turns with the rear wheel. When the brake is applied, the pedal linkage turns the cam and the brake shoes are forced outwards. The linings on the shoes grip the rotating brake drum and this slows or stops the tractor. The brake springs pull the shoes away from the drum when the pedal is released.

Each brake has an adjuster. It is used to set the shoes very close to the drum. The length of the brake pedal rods can be adjusted. When the pedals are locked together for driving on the road, it is possible that only one brake will come on if there is uneven wear on the linings. The brake pedal rods must be adjusted (balanced) to ensure that the brakes come on evenly when the pedals are locked together. Unbalanced brakes are very dangerous, especially when driving at fast speeds.

Disc Brakes

The disc brake has a greater braking surface area than an expanding shoe brake of equivalent size.

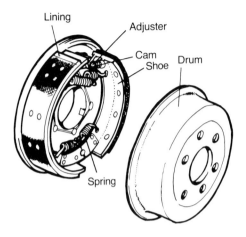

FIGURE 5.11 Expanding shoe brake.

Two rotating discs, with friction lining material on each face, are fixed on the drive shafts to the rear wheels. There are two stationary plates between the rotating discs. Two outer plates are provided by the smooth surfaces of the disc brake housing. There are some steel balls in tapered grooves between the inner stationary plates which are held together with springs. Some big tractors have three, or more, stationary plates.

As the brake pedal is pushed down, the linkage causes the inner plates to be forced apart by the steel balls. The expanding inner plates grip the brake discs against the outer stationary plates. The tractor slows down or stops completely.

The springs pull the inner plates together when the pedal is released and the rear wheels are free to turn.

Disc brakes need adjusting as the linings wear. Some tractors have an adjuster nut on each of the pedal linkages. It is not, as a general rule, necessary to balance disc brakes. There are many methods of adjusting disc brakes depending on make and model of tractor. Check with your tractor instruction manual before carrying out this important task for the first time.

In the same way that quiet cabs have brought about the use of hydraulic clutch control, some tractors have hydraulically operated brakes. Each brake pedal operates a master cylinder, which in turn applies one brake with a small ram cylinder on the disc brake assembly. A balancing valve ensures that both brake cylinders work at the same pressure when the pedals are locked together.

Many tractors have disc brakes which run in oil. The lubricant is, almost without exception, a special type with important additives to ensure trouble-free service. Wet disc brakes (those which run in oil) are usually in the main transmission housing. Always use the type of oil recommended in the instruction manual.

The handbrake
A handbrake, which over-rides the pedals when the tractor is parked, is standard equipment on most tractors. Sometimes a locking device is provided to hold the pedals down, for parking. Some of the older tractors have a handbrake on

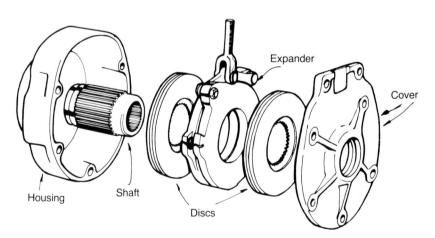

FIGURE 5.12 Disc brake.

the side of the gearbox. This is used to operate a transmission brake inside the gearbox. It is intended as a parking brake.

LUBRICATION

The lubrication of the transmission will depend upon the type of equipment used. Special oils are needed for brakes which run in oil. Hydraulic systems need the correct oil too.

Transmission oils require:

- *Anti-wear*—an additive to protect the transmission components from the very high pressures between the gear teeth.
- *Anti-oxidant*—an additive to combat the chemical breakdown of oil (oxidation) which tends to thicken the oil.
- *Brakes*—a special additive is required for transmissions which have brakes running in oil. This overcomes the problem of the brakes juddering or 'squawking' when they are used.
- *Anti-foam*—an additive to overcome the build-up of air bubbles in the oil, especially in relation to the use of hydraulic equipment.
- *Anti-corrosion*—as with all lubricants, this gives the oil the ability to combat corrosion especially from water, usually caused by condensation.

Regular oil checks and changes are important. Check with the instruction book for the change periods for your tractor. A general guide is given on page 76. It is particularly important to check transmission oil levels when the hydraulic external services are used for tipping trailers and other auxiliary rams.

FOUR-WHEEL-DRIVE

Difficult conditions such as slopes, wet ground and loose surfaces may result in a two-wheel-drive tractor being unable to work effectively. By driving the front wheels, greater traction (pulling power) can be obtained. There are two main types of four-wheel-drive layout—small front wheels and equal size wheels.

Drive to the front wheels is taken from the main gearbox by means of a transfer gearbox and forward running shaft to a housing on the front axle. The housing contains a crown wheel and pinion which carries a differential unit. The output shafts from the differential transmit power to the front wheels. These shafts have a flexible drive unit to facilitate steering and final reduction gears to give the correct shaft speed at the front wheels. An epicyclic reduction unit is commonly used for this purpose. A differential lock is usually provided in the front axle housing. This is designed to work simultaneously with the rear wheel diff-lock. A limited slip differential is used in some types of front wheel drive. It enables the front wheels

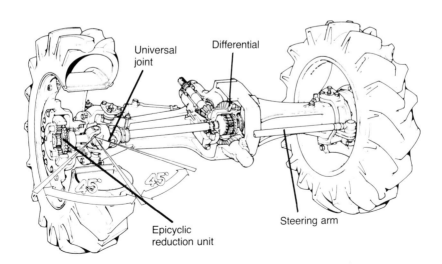

Universal joint

Differential

Epicyclic reduction unit

Steering arm

FIGURE 5.13 Line of drive to the front wheels. (*Fiatagri*)

PLATE 5.3   *A four-wheel-drive tractor with a 75 kW (100 hp) engine.*   (Massey-Ferguson)

to turn at different speeds when cornering but at other times maintains equal drive to both wheels.

On most modern tractors the drive to the front wheels is engaged by means of a switch on the instrument panel. This allows the driver to engage four-wheel-drive on the move, giving extra traction when it is required. Some early versions of four wheel drive are engaged manually with a clutch and hand lever.

Maintenance
Both transfer gearbox and front axle housing, which contains the differential unit, are lubri-

cated with gear oil. The level should be checked frequently and the oil changed at approximately 500-hour intervals. Hubs and universal joints also need regular lubrication. Check the operator handbook for the correct service intervals.

TRACTOR WHEELS

All modern tractors have pneumatic tyred wheels.

A tractor wheel consists of a dished centre disc bolted to lugs on the rim. Alternative positions for attaching the disc to the rim

PLATE 5.4 *Open centre tread tyres:* (left) *standard;* (right) *rowcrop*. (Michelin)

provide a method of varying track width to suit rowcrops etc.

Tyres for rear wheels have an open centre tread which has a self-cleaning action. When viewed from the rear, the arrow pattern of the tread must point upwards.

Tyre Sizing

The size of a tyre is marked on the tyre wall. There are three main systems of agricultural tyre sizing:

1. Tractor front wheel and implement tyres (e.g. 7.50–16). This is a tyre for a rim which is 7½ in wide and 16 in diameter.
2. Tractor driving wheels (e.g. 12.4/11–36). This is dual marking and such a tyre was originally 11–36 for an 11 in wide by 36 in diameter rim. However, when a change in rim design occurred, it altered the effective width measurement of the tyre. The 11–36 was resized to 11/12.4–36 because the measured width section had increased to 12.4 in. It is now less common to find dual sizing with the old 11–36 now known as 12.4–36.
3. Flotation tyres for trailers and implements. These tyres are used when minimum soil compaction is important (e.g. 9.0/75–18). The first figure refers to rim width. The

second is the *aspect ratio* shown as a percentage, in this case 75 per cent, and the third figure is the rim diameter.

The aspect ratio is found when the tyre section height is divided by the tyre width. For example, a tyre with a height of 9 in and a width of 12 in will have an aspect ratio as follows:

$$\frac{9}{12} = 0.75 = 75 \text{ per cent}$$

A high aspect ratio will indicate that the tyre has a very rounded section.

Ply rating refers to the number of layers of canvas or fabric used to construct the tyre. Four-ply and six-ply tyres are in common use. The heavier six-ply tyre has six layers of fabric and is necessary for the front wheels when a tractor has a front end loader fitted.

Radial and cross-ply tyres differ in the construction of the tyre wall. The radial tyre is more expensive but has a longer life with improved grip and reduced soil compaction.

Care of Tyres

Tyres must be kept at the correct working pressure. The exact pressure will depend on tyre type and size, and also on the work being done. Radial tyres will normally work at higher pressures than cross-ply tyres. Check with your tractor instruction book to find the correct pressures for your tractor.

Before the days of metrication, tyre pressures were given in pounds per square inch (psi). Tyre pressures are now given in bars or sometimes in kg/cm². One bar is equivalent to 14.7 psi; this is atmospheric pressure. A tyre with a pressure of 1 bar will also be about 15 psi. (Note: 1 kg/cm² is approximately equal to 1 bar.)

Tyre pressure
The pressure required in a tractor tyre depends on its size and the maximum load it will be expected to carry. Most tractors have radial tyres on the driving wheels. They require a higher working pressure than that for cross-ply tyres.

PLATE 5.5 *Tractor with flotation tyres.* (Colchester Tillage)

As a general guide, radial tyres require the following pressures:

Front tyres (4WD)	1.4 bar (20 psi)
Rear tyres	1.6 bar (24 psi)

When loaders or other front-mounted implements are used, the front tyres need a pressure increase of around 30 per cent.

For cross-ply tyres, typical pressures are:

Front tyres	2.0 bar (30 psi)
Rear tyres	1.0 bar (15 psi)

Cross-ply tyre pressures should be increased by about 20 per cent if the tractor is used for continuous road work. When a front-mounted loader is used, the front tyres need a pressure of about 2.7 bar (40 psi).

Full information on the correct tyre pressure for any tractor is given in the instruction manual. It indicates the pressure for a range of tyre sizes and permissible axle loadings. Exceeding the maximum recommended axle load will result in premature tyre failure.

As well as maintaining correct pressures, tyre life can also be extended by observing the following points:

1. Wipe any oil or grease from the tyres.
2. Remove stones or other sharp objects from the tread.
3. Avoid skidding and fierce use of the clutch.
4. Always replace the valve dust cap after inflating a tyre.

Tractor tyres drive partly by penetration of the soil with the tread bars or lugs and partly by adhesion (surface contact). A tyre on a car transmits drive by adhesion alone. Tyre performance can be improved on a tractor by adding weight to improve both penetration and adhesion.

High tyre pressures reduce tractive efficiency. The lowest recommended pressure should be

PLATE 5.6   *Four-wheel-drive tractor with front weights to increase stability.*   (Massey-Ferguson)

used for normal field work to obtain maximum grip.

Water ballast

Ballasting tractor tyres with water is a cheap way of adding weight. It also reduces bounce on rough land. The water must have some calcium chloride added to protect it from frost. It is usual to fill a tube about three-quarters full. This can be done without special equipment, but is a task often left for a tractor mechanic. When water ballasting is done by the tractor driver, it must be remembered to add the chemical to the water. The chemical reaction will generate heat when calcium chloride is added to water, so the solution must be allowed to cool before filling the tyre. A special adaptor is needed to fill the tube to the correct level. The valve must be at the highest position when the tube is filled. The

adaptor also allows displaced air to escape from the tube as the water is added. When the tyre is 75 per cent ballasted (three-quarters full) the water will be up to the level of the valve. The tyre is then inflated to its normal working pressure. Because the air space in the tube is very much reduced, tyre pressure must be checked more often. Use a tyre gauge which is intended for checking ballasted tyres—the calcium chloride will corrode an ordinary gauge—and rinse the gauge with clean water after use.

A 13.6/12–36 tyre, 75 per cent water ballasted will have approximately 190 kg of added weight. It is usual to mix 1 kg of chemical with every 5 litres of water.

Weights

Cast iron weights are a more expensive way of adding weight. The advantages of iron weights are that they can be removed very quickly and mending punctures is less of a problem.

Weights can be bolted to the rear wheels. They consist of cast iron discs, each weighing about 40 kg. Two or three weights are usually bolted to each wheel.

A weight can also be bolted to each front wheel or weights can be carried in a front-mounted weight tray. These are often needed to counterbalance the weight of a heavy rear-mounted implement such as a reversible plough. Front wheel weights usually weigh between 30 and 40 kg. The amount of weight carried in a weight tray depends mainly on size of tractor. A typical 45 kW (60 hp) tractor may have approximately 200 kg in the weight tray.

Weight transfer

The tractor hydraulic system can be used to transfer some of the implement weight on to the tractor to improve wheel grip.

The performance of tractor tyres can also be improved by using strakes. These are bolted on to the inside or outside of the rear wheels, depending on type. The strake has spikes which can be extended beyond the tyre. The spikes dig into the soil to improve wheel grip. Tractor performance can also be improved with the diff-lock, especially in wet conditions where wheelspin is a serious problem.

Most tractors can have an extra pair of rear wheels, usually attached with a quick fit or release mechanism. Dual wheels serve two purposes. They can be used for seedbed work to reduce soil compaction, and the increased 'footprint' area keeps the tractor on the surface when working on soft or marshy land. A tractor, forage harvester and trailer could all have dual wheels when forage harvesting on marshland. Large four-wheel-drive tractors often have dual wheels on both axles as standard equipment.

Cage wheels consist of tubular or angle iron tread bars welded on to two large metal rings which form the wheel rim. They have a quick release linkage which holds them on to lugs attached to the rear wheel rims. Cage wheels reduce soil compaction because the larger footprint area distributes tractor weight to give a lower ground pressure. They are widely used for drilling and seedbed work. Cage wheels are

PLATE 5.7 *Power adjusted rear wheel. The wheel centre is rotated by engine power on the spiral rails which form part of the wheel rim. The rail near the tube valve has a series of equally spaced holes, used in conjunction with a stop, to select the required track setting.* (Massey-Ferguson)

normally used with pneumatic tyres but can be used on their own.

Wheel Track Settings

Maximum and minimum wheel track depends on size of tractor. Rowcrop models usually have track settings from 1.32–1.93 m (52–76 in). Larger models will have a minimum track of 1.42 m (56 in) and a maximum setting of 2.13 m (84 in). Adjustment is in 100 mm (4 in) steps.

For example, a crop planted at 500 mm (20 in) row spacing will require a wheel setting of 1.5 m (60 in).

Rear wheel track settings are altered by changing the position of the wheel centres in relation to the rim lugs. The wheel discs can also be reversed to vary the track setting.

Rear tractor wheels are heavy, especially when ballasted. Make sure you have the proper equipment and have help available when removing or fitting a rear wheel. Make sure the axle has a stand or firm wooden block under it, to take the weight of the tractor before removing the wheel. Remember to slacken the wheel nuts half a turn before taking the weight off the tyre. Chock the other wheels with blocks. When replacing the rear wheel take care not to damage the threads on the studs. Fit the top nut first. When all the nuts are finger tight, tighten the top one first and then the remaining ones in diagonal pairs. Carry out final tightening after lowering the wheel to the ground.

Power adjusted rear wheels are fitted to some tractors, especially the larger models. The rim is like a giant nut and the wheel disc is the bolt. The wheel track can be adjusted by rotating the disc in the rim using engine power and selecting either a low forward or reverse gear depending on which way the rim is to be moved. The wheel being adjusted is allowed to turn; the other is locked with the foot brake. Clamps which secure the disc in the required position in the rim are removed before making the adjustment. Once adjusted, the clamps must be replaced and secured.

Two-wheel-drive tractors have a telescopic front axle design. The front track setting is altered by removing the axle beam bolts and then sliding the movable section inwards or outwards

Disc/rim positions

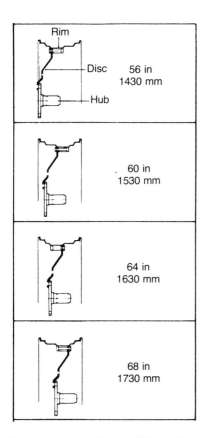
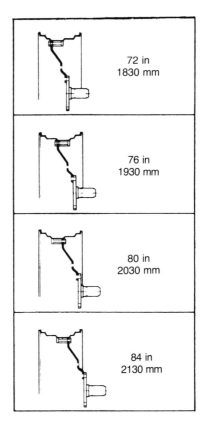

FIGURE 5.14 Rear wheel track settings. (*Massey-Ferguson*)

PLATE 5.8   *Front wheel track setting. The telescopic front axle and link from the steering ram are extended to give a wide wheel track.*   (Massey-Ferguson)

depending on the required setting. Here again, the jacked-up tractor must be secured with axle stands before starting work. Very wide track settings are obtained by reversing the wheel on its hub so that it is dished outwards. It will be necessary to adjust the toe-in (see page 74) on most tractors after completion of track adjustment. The front track setting is measured at ground level.

Front track setting on four-wheel-drive tractors is adjusted in the same way as rear wheels, by changing the position of the wheel disc in relation to the rim. A locking device is used when the front wheel tracks are adjusted to the wider settings; this prevents the tyres fouling the tractor.

SUGGESTED STUDENT ACTIVITIES

1. Check the clutch pedal free play on your tractor. The correct setting is given in the instruction book.
2. Check your brakes often. The tractor should pull up squarely when the independent pedals

PLATE 5.9 *Front wheel track setting. The front axle and link from the steering ram are set for a 1,330 mm track. One hole, still visible, is left to reduce the track by a further 100 mm (50 mm on each axle).*

(Massey-Ferguson)

are locked together; if they do not, make sure that the fault is corrected.

3. Find out how the wheel track setting is adjusted on different models of tractor.
4. Look at the markings on tractor tyres and interpret their meaning.
5. Check, at regular intervals, the tyre pressures of your tractor.

SAFETY CHECK

Never engage the drive to a tractor power take-off shaft until you have checked that all guards are in place.

Ensure that independent brakes are locked together when travelling along public highways.

Chapter 6

TRACTOR HYDRAULIC SYSTEMS

The tractor hydraulic system provides a means of raising and lowering implements and machines into and out of work. Most implements are carried on the rear linkage system, but with the popularity of larger tractors a front linkage is often employed to make full use of the available tractor power and carry out more than one field operation at a time. Some examples include a rear-mounted plough with a front-mounted furrow press, and a front-mounted cultivator with a drill on the rear linkage.

The system also provides power to operate external rams on a wide range of machines and the oil flow required by hydraulic motors used on a variety of implements such as hedge cutters and root harvesters.

The Principles of Hydraulics

Liquids can be used to transmit power from one point to another. In a simple hydraulic system, a pump supplies oil at a high pressure to a ram cylinder. The oil forces the piston (ram) along in its cylinder. The movement of the piston is used

PLATE 6.1 *The tractor hydraulic system providing the lifting power to carry three bales, one on the front-end loader and two on the rear mounted transporter.* (Brown)

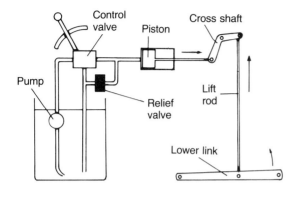

FIGURE 6.1   Principles of hydraulic lift system.

to lift a load. A hand-operated hydraulic jack has a hand pump which forces oil against the piston in its cylinder. The piston moves upwards and lifts the vehicle.

Figure 6.1 shows a very basic hydraulic lift system. Oil is pumped through the control valve to the ram cylinder.

The control valve is set in the *lift* position. The oil forces the piston along the cylinder. The piston is connected to a ram cross-shaft by a very simple connecting rod. Movement of the piston turns the ram cross-shaft on its pivots. This raises the lower link arms by means of the lift rods.

Oil is released from the ram cylinder back to the reservoir, usually the transmission housing, when the control valve is moved to the *lower* position. The implement falls under its own weight, forcing the oil from the ram cylinder. A relief valve allows oil to return to the reservoir if the system is overloaded.

This simple type of hydraulic system requires a wheel to control the working height or depth of the implement. Modern hydraulic systems control working depth automatically, making a depth wheel unnecessary.

Pumps
A gear pump is commonly used for tractor hydraulics. Some models have a multi-piston pump. Most pumps are built round the power take-off shaft, inside the transmission housing. Some tractors have an engine-mounted pump which is quite independent of the transmission

system. A typical 45 kW (60 hp) tractor has a pump with an output of 27 litres of oil per minute.

Filters
The hydraulic oil filter may be no more than a magnetic plug to catch any minute particles of metal in the oil. Some tractors have a replaceable element-type filter for the transmission oil. The oil must be clean, to prevent damage to the pump and control system.

Relief valve
This prevents damage to the pump through overloading. It does this by limiting the pump output pressure. The relief valve will 'blow' (come into operation) if the driver attempts to lift too much on the lift arms of a front-end loader. The pump output pressure increases as the load increases. When the pressure reaches the maximum allowed by the relief valve, it will 'blow'. A knocking sound in the transmission housing can often be heard when the relief valve comes into operation.

A typical 45 kW tractor will have a 'blow-off' pressure of 150 kg/cm² (2,000 psi).

CONTROL SYSTEMS

Tractor hydraulics have a number of different systems to suit various types of work. The actual control levers vary from one tractor to another. The more important hydraulic control systems are described below.

Draft Control

Draft control is used for implements which work in the soil. Ploughs, cultivators and subsoilers are examples. The draft control system maintains a constant load (draft) on the tractor. In ideal working conditions, with the soil type and conditions the same all over the field, the load on the tractor and therefore the working depth would be constant. In practice, of course, this does not happen.

The tractor driver can use draft control to set his plough at a chosen depth. The hydraulic system will have a certain load to pull at this depth. Draft control will keep this load at a constant level and, at the same time, give an

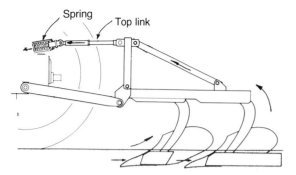

FIGURE 6.2   How top link sensing works.

acceptable standard of depth control. Very heavy mounted implements, especially ploughs, often need a wheel to help maintain the required depth.

Top link sensing

A mounted implement is carried by the lower links. The top link stops the implement tipping backwards or forwards. When a soil-engaging implement is at work, there is a force acting along the top link towards the tractor. This force becomes greater as the working depth increases.

The draft control system uses the forces in the top link to maintain a constant load on the tractor. There is a linkage connecting the control valve to the top link. When the implement drops below the required working depth, the force in the top link increases. The connecting linkage moves the control valve slightly and the implement is raised to its correct working depth.

There is a heavy spring at the end of the top link. This is sometimes visible but many tractors have the spring inside the transmission housing top cover. This spring must be compressed by the force acting in the top link before any correction to working depth can be made. The spring is also used to push the top link in the opposite direction when the force from the implement is reduced. This will happen if the implement is working above the required depth. The spring moves the control valve linkage which allows the implement to drop back to the correct depth.

In very uneven soil conditions there would be too many depth corrections. To overcome this, most hydraulic systems have a damping or response control, with a range of settings from fast to slow. When depth control operation is sluggish, the response control is set in a fast position. A slow response setting is required if the implement is making too many corrections to its working depth. The draft control system already described only uses changes in the compression force acting in the top link. The weight of an implement working at a shallow depth can create a tension force in the top link which is greater than the compression force produced by the implement working in the soil.

Most tractors have a double-acting top link. This will transmit both compression and tension force signals to the hydraulic control valve, to maintain a constant draft (or depth). The top link will be in tension when shallow ploughing. There will be an increase in the strength of the tension force in the top link if the tractor front wheels drop into a furrow. The tension in the top link will increase and the control valve will return the plough to the pre-set position. The top link is always in tension when transporting mounted implements. To prevent damage to the hydraulic system, especially when moving over rough ground, a mechanism cuts out the effect of tension forces when the implement is fully raised.

Lower link sensing

The tension force acting in the lower links can also be used to give draft control. This is called lower link sensing. It relies on the fact that as an implement goes deeper, the draft and the tension force in the lower links will increase. A sensing mechanism connects the lower link attachment points to the hydraulic control valve. Increased draft on the lower links will cause the control valve to lift the implement slightly. In the same way, the implement will be lowered when draft decreases.

Constant implement draft (depth) can also be maintained by using the changes which occur in the torque (twisting force) in the transmission system. Overload, due to increased working depth or heavier soil conditions, will increase the driving shaft torque. Some tractors have a special coupling built into the drive shaft from the gearbox to the crown wheel and pinion, which reacts to variations in torque. The coupling is connected by a linkage to the hydraulic control valve.

When the tractor is overloaded, beyond the pre-set draft, the coupling reacts and the linkage moves the control valve. The implement is lifted

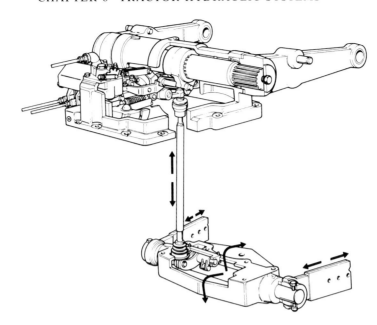

FIGURE 6.3   The principle of lower link sensing. Increased tension, through extra loading, causes the lower links to raise the plough slightly.   (*Fiatagri*)

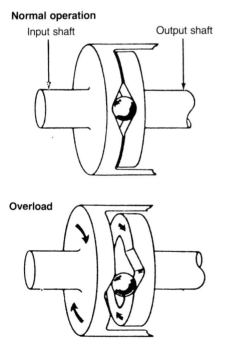

Normal operation

Input shaft

Output shaft

Overload

FIGURE 6.4   When draft exceeds pre-set level, the input shaft turns in relation to the output. This expands the torque coupling, this movement is transmitted to the hydraulic valve and the implement is then lifted to restore the pre-set level of draft.   (*Ford*)

slightly, returning it to its pre-set draft position. This system will also increase the implement depth when draft is reduced.

Semi-mounted implements can be used to advantage with this type of draft control. It can monitor changes, for example in the load on a tractor pulling a semi-mounted plough when depth or soil texture change. The control unit operates the auxiliary ram at the back of the plough to increase or reduce the depth of the rear bodies while the lower links control the depth of the front bodies. This degree of control cannot be achieved with normal top link sensing because the semi-mounted implement is carried by wheels at the rear.

Electronic control systems
Corrections to changes in draft in the hydraulic linkage, caused by variations in soil conditions, can be made electronically on certain models of tractor. Special sensing pins in the linkage measure changes in the draft load and the position of the lower links. The sensing pins relay instructions to the control system which automatically readjusts the depth, and draft, of the implement.

The same electronic control system also enables the driver to control rate of implement drop and the maximum height of implement lift. It can also lock out the hydraulic system, making it safe when the tractor engine is stopped.

Some top of the range tractors, with in-cab computers, have a further refinement which gives more accurate control of draft in the hydraulic linkage. The system is pre-set to allow a certain percentage of wheelslip. The normal acceptable level of wheelslip is about 15 per cent. When wheelspin exceeds the pre-set level, the implement is automatically lifted slightly. As it lifts, wheelspin is reduced because extra implement weight is transferred on to the tractor's rear wheels, which improves traction. The implement returns to the pre-set depth immediately the excess wheelspin has been eliminated. This process may be repeated several times each minute when working in difficult soil conditions.

Without computerised wheelslip control, the driver could obtain a similar effect by lifting the implement slightly with the draft control lever when wheelspin occurs. However, fast and accurate manual control would be very difficult to achieve and, as sometimes happens in difficult

working conditions, if the driver is not quick enough on the lever, the tractor gets stuck.

Position Control

This system is used for mounted implements which work above ground level such as sprayers, fertiliser spreaders and mowers. The position control lever is used to set the required working height on the three-point linkage. Any changes in the forces acting in the top link will have no effect on the control valve.

Flow Control

The rate of oil flow to and from the ram cylinder can be altered with the flow control valve. It is used to control either rate of lift or drop depending on model of tractor.

Mechanical Lock

Some tractors have a mechanical lock which prevents the implement being lowered when it is engaged. Operated with a hand lever, the locking mechanism removes the risk of damage to the hydraulic system when transporting heavy implements, especially when travelling at speed on the highway.

Auxiliary Hydraulic Services

The tractor hydraulic system is required to supply oil to other hydraulic services as well as to the three point linkage. Many machines have auxiliary rams. Hydraulic motors also rely on the tractor auxiliary service system.

The most common use of auxiliary services is the tipping trailer, which has a ram supplied with oil from the tractor hydraulic pump. A single-acting ram is used on a tipping trailer. Oil is pumped to the ram for lifting the trailer, but the weight of the trailer body is needed to return the oil to the tractor as the ram is closed.

Many other implements have rams, some of them double acting. This type has the ability to operate in two directions. It has one hose pipe to supply oil to the ram and a second to return it from the ram to the tractor. Oil can flow in either direction in both pipes depending on the position of the control lever. A reversible plough is a typical example of the use of a double-acting ram.

Remote control valves are used to operate auxiliary rams. They can be used to control either single- or double-acting rams. The hoses have quick release connectors which are plugged into the hydraulic system. Tractors may have as many as eight auxiliary service tapping points, allowing four separate double-acting rams to be connected to the system. Levers are provided to control the oil flow to and from the rams.

Earlier models of tractor have hose connections which have screw-type couplings. The control lever is built into the main hydraulic control levers, which is combined with an isolator tap on some tractors.

The auxiliary hydraulic service also provides oil flow for hydraulic motors. Two hose connections are required, one for flow and one for return. The advantages of hydraulic motor drive are that it eliminates expensive drive shafts which need guarding and oil flow to the motor can be regulated to give variable speeds in forward or reverse at the touch of a lever.

PLATE 6.2 *Hydraulic linkage and auxiliary ram couplings.* (Massey-Ferguson)

The Hydraulic Linkage

The lower links are connected to the ram cross-shaft by the lift rods. The right-hand lift rod has a levelling box which is used to adjust the height of the right-hand lower link.

The top link, usually adjustable in length, is used to adjust the pitch (angle) of some implements such as ploughs, cultivators, sprayers and fertiliser broadcasters.

Stabiliser bars are used to hold a mounted implement rigid on the tractor. Some tractors have telescopic stabilisers which can be set in a rigid position or adjusted to allow some sideways movement of the implement.

External check chains have the same purpose as stabilisers and also offer limited sideways movement where required. Some tractor manufacturers prefer check chains to stabiliser bars.

Internal check chains allow free sideways movement but prevent the lower links striking the rear tyres when the implement swings on the linkage.

The linkage hitch pins are made in three different sizes. They vary in diameter and length. The measurement between the three hitch point positions on the implement are also different.

Category One is the smallest size.

Category Two is the middle size linkage, which is in common use. Many tractors have dual linkage with both categories one and two. This is achieved by changing the hitch balls in the lower links and top link.

Category Three is the heavy duty linkage used on the largest tractors with very heavy implements.

Remember to attach mounted implements in the correct order:

1. Lower left-hand link.
2. Lower right-hand link.
3. The top link.

A quick attach linkage system is available as an alternative to the standard ball hitch system. The lower link balls are replaced by telescopic hook ends which have self-locking latches. The telescopic ends allow for inaccuracies when reversing up to the implement. Ball-shaped brushes with conical guides are fitted to the implement hitch pins. The implement is attached by reversing up to the implement with the hooks under the

PLATE 6.3 *Quick attachment link arm ends.* (Massey-Ferguson)

attachment point. When the lower links are raised, the hooks engage and lock. The top link hook end is engaged with the implement and locked in position. When the implement is raised, the telescopic ends lock in the lower links.

Uncoupling is achieved by releasing the hook end locks with a release cable which is within reach of the driving seat. The top link is lifted clear and the lower links are dropped to their lowest position.

Care of the Hydraulic System

1. Grease the lift linkage and the levelling box weekly.
2. Check transmission oil level – or separate oil reservoir – every week. This is especially important if a tipping trailer or similar external ram has been used and oil has been lost during coupling or uncoupling the pipes.
3. Do *not* lubricate the balls in the lower links. They will collect grit and this will cause rapid wear.
4. Never turn a sharp corner with the implement still in the soil. This puts unnecessary strain on the linkage.
5. Always make sure a mounted implement is lowered to the ground before leaving a parked tractor.

SUGGESTED STUDENT ACTIVITIES

1. Read the section in your tractor instruction book which deals with hydraulic controls.
2. Look at the methods used on various tractors to convert from one category of linkage to another.
3. Find out what type of draft control sensing is used on your tractor.
4. Look for examples of stabilisers and check chains on different tractors.
5. Practise reversing up to a mounted implement until you can accurately line up the lower links with the hitch pins.

SAFETY CHECK

Never tow anything from the top link position. This is a dangerous practice because it can result in the tractor somersaulting backwards. At a forward speed of 3.25 km/h (2 mph) a tractor will rear up through an angle of 90 degrees in one second.

Chapter 7

TRACTOR STEERING

The farm tractor can have manual, power-assisted or hydrostatic steering. Power-assisted steering is of considerable advantage when using a front-mounted loader. Hydrostatic steering has been used for combine harvesters for some while, but the introduction of quiet cabs has brought this type of steering into increasingly common use for tractors too.

PLATE 7.1 *An adjustable steering wheel in a luxury cab with temperature control and all round vision through tinted glass.* (Massey-Ferguson)

Manual Steering Systems

The steering wheel is used to move the drop arm backwards and forwards by means of the steering gearbox. The drop arm is connected to the steering arm by the drag link. When the drop arm moves backwards, the drag link pulls the steering arm round and changes the direction in which the tractor travels. There is a steering arm for each wheel; they are connected with the track rod. This gives an equal amount of movement to both steering arms.

The track rod length must be reset after changing the front wheel track width.

Another type of manual steering has two drop arms operated by the steering box. When the steering wheel is turned, one drop arm moves forward and the other moves backwards. The drop arms are connected to the steering arms by drag links. There is no track rod. The action of the drop arms and drag links turns both wheels in the same direction when the steering wheel is moved.

Some tractors need an adjustment to drag link length after altering wheel track widths; others do not.

Power-assisted Steering

With this type of steering, there is a manual linkage with a miniature hydraulic system to help move the steering arms to the left or right. The hydraulic system consists of a pump, an oil reservoir and a ram cylinder. The pump is usually fan belt driven. The control valve is operated by a linkage from the steering box.

Tractors with power-assisted steering can still be steered manually if the hydraulic system breaks down. The ram is double-acting. This means that oil can be pumped to either side of the ram piston and give movement in two directions. When turning to the left, the steering mechanism operates the control valve to direct oil to the side of the piston which will move the front wheels to the left. Oil from the other side of the piston is released back to the reservoir.

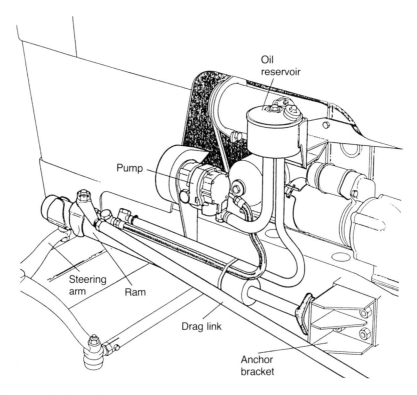

FIGURE 7.1 Power-assisted steering.

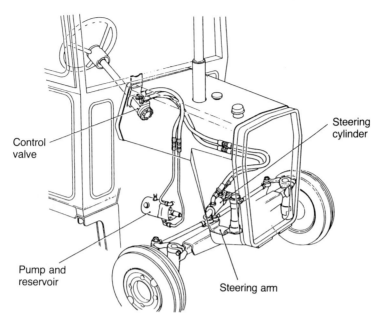

Control
valve

Steering
cylinder

Pump and
reservoir

Steering arm

FIGURE 7.2   Hydrostatic steering.

Hydrostatic Steering

There is no conventional steering linkage from the steering wheel to the front wheels. The main components are an engine-driven pump, a double-acting ram, an oil reservoir and piping. The control valve is operated by the steering wheel. The tractor can only be steered when the engine is running. It is not possible to steer the tractor manually if the hydrostatic system develops a fault.

The control valve directs oil to one side of the ram piston. Oil is released from the other side of the piston back to the reservoir. When the driver wishes to turn left, the steering wheel moves the control valve. Oil is directed to the side of the ram cylinder which, through a linkage, turns the front wheels to the left.

Steering Boxes

Tractors with manual steering systems usually have a recirculating ball gearbox. It converts the rotary motion of the steering wheel into linear movement which, by means of a linkage, turns the wheels in the required direction.

A variation of the recirculating ball gearbox, which includes a control valve unit to operate the steering ram, is used on many power-assisted

steering systems. See Figure 7.3 to follow the operation of the recirculating ball gearbox. The end of the steering column has a thread (A). A nut (B) moves on the thread when the steering wheel turns. This nut is connected to a rocker arm (D) by a peg (F). Steel balls run in the grooves formed by the threads on the steering shaft and the nut; this gives smooth movement of the steering mechanism. When the nut (B) moves along the thread, the rocker arm and shaft (E) rotate. The drop arm is fixed to this shaft. As the shaft rotates through about 30 degrees backwards or

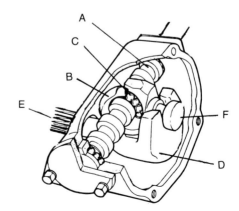

FIGURE 7.3   Recirculating ball steering gearbox.

forwards, it pulls or pushes on the drag link to change the direction of travel.

A second type of recirculating ball steering box is shown in Figure 7.4. This box has two rocker shafts and is used with a double drop arm system. When the wheel is turned, the two rocker shafts rotate in opposite directions. This causes one drag link to move forwards and the other to move rearwards so that both front wheels turn in the same direction. This illustration also shows the valve assembly which controls the oil flow to the hydraulic ram.

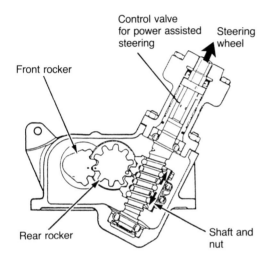

FIGURE 7.4   Steering gearbox for double drop arm linkage. (*Ford*)

Maintenance
As a general rule, the steering linkage should be greased every 10 working hours. Some manufacturers suggest less frequent lubrication, so check with your tractor manual. The grease nipples are usually to be found on the steering arms, the stub axles, and also on the track rod and drag link ends. More use is being made of sealed bearings; this has reduced the number of lubrication points on many steering linkages.

Some tractor steering boxes have a level plug. This should be removed every 200 hours to check the oil level. SAE 90 is commonly used for steering boxes.

Front wheel bearings should be lubricated every 50 hours. Some wheels have grease nipples;

others require the removal of the hub caps which are then packed with grease. From time to time, check for front wheel bearing wear. To do this, first jack the tractor front wheels clear of the ground. Check to see if it is possible to rock the wheels on the stub axles. Place one hand at the top of the wheel and the other hand at the bottom and try to move the wheel on its bearings. Rocking movement shows that readjustment is required. Remove the hub cap and take the split pin (if fitted) from the bearing end nut. Tighten the nut until the slack has been removed but make sure the wheel still turns freely. Fit a new split pin, and after packing the hub cap with grease, replace it on the wheel.

Toe-in

Front wheels are said to have toe-in when the measurement between the rims is less at the front than it is at the back. The measurement is taken at wheel bearing height. Toe-in improves steering performance and reduces front tyre wear.

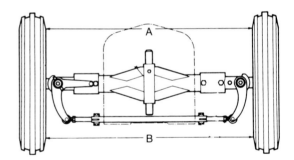

FIGURE 7.5   Setting for toe-in.

In Figure 7.5, distance A should be from 3 to 6 mm less than distance B. This is a typical setting but the measurement will vary with different models of tractor. Toe-in is altered by a small adjustment to the length of the track rod. The track rod ends are threaded into the tubular centre section and locked in position with clamp bolts. Some four-wheel-drive tractors do not have toe-in because the front axle is power driven.

Four-wheel-drive steering
Tractors with four-wheel-drive will have either power-assisted or hydrostatic steering. Since the

front axle is driven, the normal two-wheel-drive steering linkage is not suitable. Universal drive couplings in the drive shafts enable the wheels to change direction while full power is delivered to the front wheels.

SUGGESTED STUDENT ACTIVITIES

1. Look for and compare the different tractor steering systems.
2. Locate all grease nipples on the steering linkage. Check with the instruction book to find the correct lubrication periods.

3. Check the front wheel bearings from time to time (two-wheel-drive tractors). Adjust them if necessary.
4. Locate the steering box and power steering oil reservoir. Check the oil levels and top up with the correct grade of oil if necessary.

SAFETY CHECK

Always place an axle stand under a tractor axle before removing a wheel. Chock the wheels remaining on the ground and do not forget to apply the handbrake too.

Chapter 8

TRACTOR MAINTENANCE

The regular maintenance needs of farm tractors vary from one model to another, both in service periods and range of tasks. The operator's manual sets out the correct service schedule. You should refer to this schedule before carrying out routine maintenance to your tractor.

Remember that the periodic service hours refer to the *engine* hours shown on the proof meter. In most cases, the proof meter registers the hours the engine has run at about two-thirds maximum speed. So, in a normal day's work, the proof meter will register at least ten hours. However, some of the latest tractors record

PLATE 8.1 *Engine servicing. Can you locate the oil filter, the diesel fuel filter and sediment trap, the dipstick and the oil filler in this picture?*

(Massey-Ferguson)

actual hours worked; they are not related to engine speed.

Typical Service Schedule

10 hours (or daily)

- Check engine oil level and top up with correct grade if necessary.
- Check radiator water level, and top up if required. Remember to use a mixture of anti-freeze and water in the winter.
- Lubricate steering linkage. (*Note*: some instruction books suggest this task should be done at intervals of 50 hours.)
- In dusty working conditions, check the oil in the air cleaner oil bath and service if necessary. Remove any chaff or other rubbish from the gauze filter. For tractors fitted with a dry air cleaner, check the element and clean if necessary.
- Make a visual check of tyres and test air pressure if considered necessary.
- Fill windscreen washer reservoir if fitted.

50 hours (or weekly)
Carry out 10-hour service tasks and:

- Check oil levels in gearbox and transmission and top up if necessary.
- Lubricate wheel bearings, clutch and brake linkages and the hydraulic linkage levelling box.
- Service the air cleaner.
- Check tyre pressures and inflate if necessary.
- Check the battery and top up with distilled water if required. Keep the battery terminals clean and tight.
- Check wheel nuts and front axle bolts for tightness.
- Top up level of fluid in the power-assisted or

LUBRICATION POINTS

GREASE EVERY 10 HOURS, OR DAILY

1. Front hubs
2. Front axle kingpins
3. Front axle pivot pin
4. Main steering shaft
5. Power steering ram pivot
6. Lift rods
7. Levelling box
8. Hook links
9. Auto-hitch latches

FOUR WHEEL DRIVE

10. Front support pivot
11. Drive shaft couplers front and rear
12. Rear support input housing right-hand side
13. Drive shaft centre bearing
    Grease every 100 hours or in wet or muddy conditions
    every 10 hours or daily.

EVERY 100 HOURS

Apply a smear of light oil to the cab door hinges and locks,
hood side panel hinges and latches, and front grille latch.

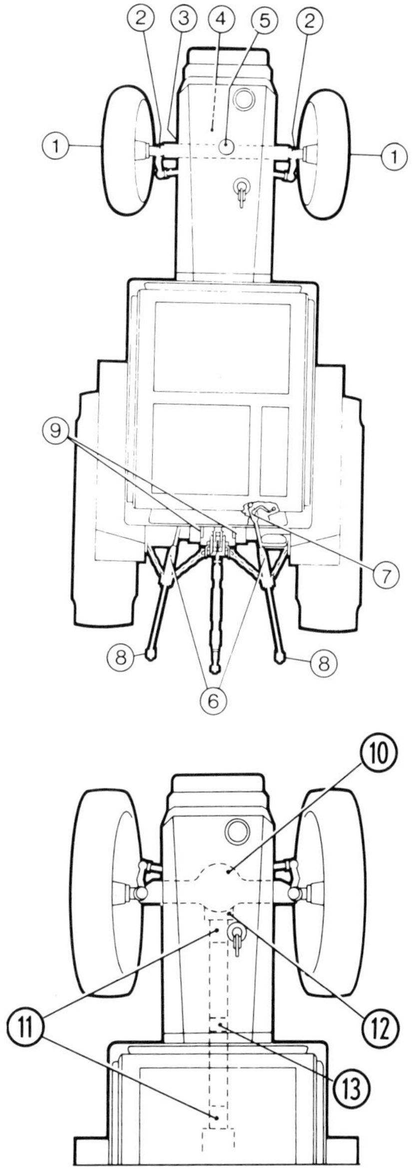

FIGURE 8.1   A page from a tractor service schedule.   (*Massey-Ferguson*)

PLATE 8.2 *Removing the oil bath from an air cleaner. The hose connection* (top left) *which supplies clean air to the engine must be airtight.*

(Massey-Ferguson)

hydrostatic steering boxes, also the clutch and brake fluid reservoirs. Use the correct grade of fluid.

300 hours
Complete 10- and 50-hour service and:

- Drain the oil from the engine when it is hot. Renew filter element and refill the sump with fresh oil. *Note*: the engine oil change period varies according to tractor and grade of oil used.
- Clean the crankcase breather.
- Check steering box oil level and top up if necessary.
- Make sure the clutch pedal free play is correctly adjusted. Reset if necessary.
- Check brakes for adjustment and, where appropriate, balance.
- Drain any water collected in the diesel fuel filter water trap.
- Check fan belt tension and adjust if required.
- Make sure the radiator core is clean externally. Remove chaff or dust with an airline or hose pipe.

Note: turbocharged engines require more frequent engine oil changes. The correct specification and grade of oil must be used.

600 hours
In addition to all the tasks listed above:

- Renew the diesel fuel filter element and bleed the fuel system.
- Arrange for the diesel fuel injectors to be checked and if necessary have them replaced.
- Change the transmission system oil. This task is more often advised after 1,000 hours.

You should refer to the tractor handbook to find the servicing requirements for specialist equipment such as front-wheel-drive and hydrostatic steering.

The strength of the anti-freeze in the cooling system should be checked and topped up in readiness for the winter months.

Chapter 9

PLOUGHS

The plough has been used in its different forms for many centuries. It is still the primary cultivation tool used in the production of seedbeds. Good quality ploughing speeds up the preparation of seedbeds at any time of the year.

The plough is an implement with one or more mouldboards which cut and turn furrow slices. Modern ploughs are either mounted or semi-mounted. Three-, four- and five-furrow ploughs are the most popular, but models with from one to twelve furrows are available. Semi-mounted ploughs are carried on the tractor hydraulic linkage at the front and supported by wheels at the rear. They are never lifted completely clear of the ground.

A push-pull arrangement with one plough

PLATE 9.1   *Reversible ploughing. Note the tilth made by the plough and furrow press combination. The furrow press drawbar can be seen behind the plough, ready for the next bout.*   (Colchester Tillage)

79

PLATE 9.2   *Four-wheel-drive tractor with a push-pull plough. This implement requires careful matching of the front and rear ploughs in order to achieve high quality work.*   (Agrolux)

on the rear linkage and a second carried on a front linkage is an alternative design. Various combinations are available but three furrows at the front and four at the back is a typical arrangement for the larger tractor.

There are still a few trailed ploughs in use with crawler tractors. A visit to a ploughing match will reveal many competitors using trailed ploughs, mainly with vintage and veteran tractors. The work done by the competitors sets a standard which should be the aim of all ploughmen.

Most ploughs on British farms are the reversible type which owe their origin to the days of steam ploughing. However, in many countries, the conventional single-handed plough is in common use and most plough manufacturers offer a full range of this type of plough.

PLOUGHS

There are three main types of plough:

Conventional ploughs
These have right-handed mouldboards. They are either mounted or semi-mounted and may have one to eight furrows. Some manufacturers produce trailed models with twelve furrows. Conventional ploughs are much lighter than reversible ploughs, and also much cheaper.

Reversible ploughs
These have left- and right-handed mouldboards, enabling the plough to work up and down the same furrow. This type is either mounted or semi-mounted. They are much heavier and more

PLATE 9.3   *A seven-furrow semi-mounted conventional plough. It has a ram at the back of the plough to lift the rear bodies out of work. The front of the plough is raised by the hydraulic lift arms. The rear wheel of this plough is steerable.*   (Dowdeswell)

expensive than conventional ploughs but have the advantage of leaving a level field, making seedbed preparations and harvesting easier. Very little marking out is necessary before ploughing can start and idle running on the headland is minimal compared with using a conventional plough.

Disc ploughs

Rarely seen in Great Britain. They have large rotating discs instead of mouldboards. These discs cut and turn the furrow slice but do not bury the trash. Disc ploughs are in common use in hot countries where failure to bury trash efficiently does not matter, as it is quickly scorched by the sun. Both conventional and reversible disc ploughs are made.

The Parts of a Plough

The soil engaging parts, disc coulter, skim coulter and the body are attached to legs which are in turn bolted to the plough beam or frame.

The base of a plough body is called the *frog*. The soil engaging parts are bolted to it.

The *share* cuts the bottom of the furrow slice.

The *mouldboard* lifts and turns the furrow slice. There are many different types of mouldboard, each producing its own surface finish.

The *landslide* absorbs the side thrust of the plough against the furrow wall.

The *heel iron*, bolted to the rear landslide, helps to carry the weight of the plough at the back.

The *mouldboard extension* helps to press down the furrow slice.

PLATE 9.4   *A mounted reversible plough at work. This model has a ram and linkage which is used to vary the width of all the furrows, even when the plough is in the ground.*   (Kverneland)

The *disc coulter* cuts the side of the furrow about to be turned. A knife coulter serves the same purpose. It is like a large knife, pulled through the soil. It takes up very little space but leaves a less defined furrow wall.

The *skim coulter* turns a small slice off the corner of the furrow about to be turned and throws it into the previous furrow bottom. This reduces the risk of rubbish growing up between the furrows.

Body Types

There are many designs of plough body, each having its particular use. The main types are:

General purpose   A low draft body with a gently curved mouldboard. It turns a furrow three parts wide by two parts deep, e.g. 300 mm wide by 200 mm deep. It is useful for grassland ploughing and sets the land up for weathering by winter frosts.

Semi-digger   This has a shorter mouldboard with a more abrupt curve. It turns an almost square furrow and leaves a more broken finish. The semi-digger body is used for most of the general ploughing on the farm.

Digger   This body has a short, abruptly curved mouldboard. It turns a furrow deeper than its width and leaves a very broken surface. Used

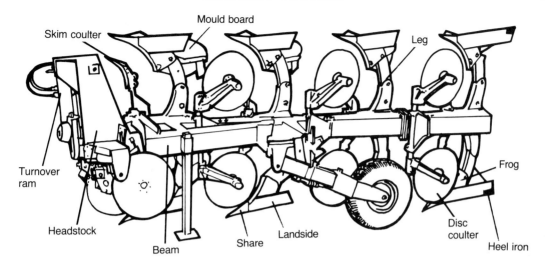

FIGURE 9.1   The parts of a reversible plough.   (*Agrolux*)

for deep ploughing, especially in preparation for root crops, the digger body has a higher power requirement than the other types.

In theory, most farms need ploughs with at least two different body types, probably general purpose and digger. Many ploughs, both those imported from Europe and those made in Great Britain, have a body with a mouldboard of continental design. This type of mouldboard will produce good work at a range of working depths. The ploughing speed will decrease as the ploughing depth increases. The popularity of the more traditional mouldboard shapes has declined in recent years since the introduction of the continental pattern.

Slatted mouldboards are preferred by some farmers. They consist of a number of curved slats bolted to the frog which are designed to improve soil movement when working in sticky conditions.

Bar point A special type of share which can be used with most types of mouldboard. An extendable bar, often spring loaded, can be moved forward as the point wears. The bar point share is ideal for soil conditions where boulders occur close to the surface. The spring-loaded bar point can recoil when it strikes an obstruction below the surface. After passing the obstruction, the bar returns to its working position. Ordinary shares would be damaged in these conditions so,

although the bar point leaves a rather untidy furrow bottom, it saves in the cost of share replacement.

Automatic reset Most plough manufacturers offer an automatic reset system for use in rocky or tough soil conditions. The reset mechanism allows each body to swing back, out of work, if it hits an obstruction. A heavy leaf or coil spring, which holds the body in its working position under normal conditions, resets the body after the obstruction is passed.

Variable furrow width
Most ploughs have a fixed furrow width, usually 250, 300, 350 or 400 mm (10, 12, 14 or 16 in). The plough will be set at one of these widths but can be altered with a toolkit; this is a lengthy process, though, and is rarely done in practice. There are times when it is desirable to adjust the furrow width. For example, in good conditions the tractor may have enough power to pull wider furrows, thus increasing the work rate. Variation in furrow width may also be desirable when there is a need to change ploughing depth, improve burial of trash, or drive at a higher speed to improve the crumbling effect of the mouldboards.

Some models of plough, both conventional and reversible, have an adjustment which enables the driver to change furrow width, either with a turnbuckle or a hydraulic ram. The body units

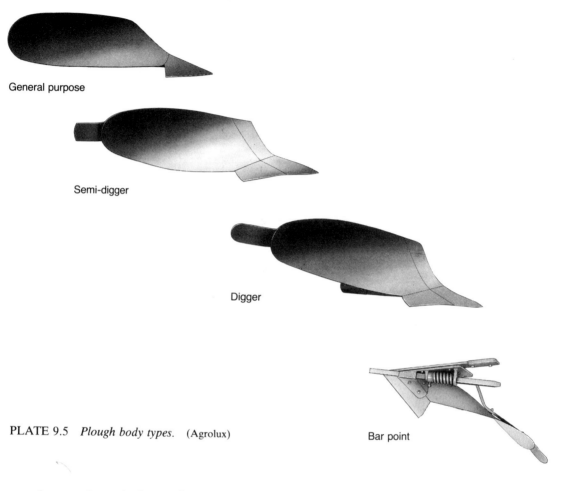

General purpose

Semi-digger

Digger

Bar point

PLATE 9.5   *Plough body types.*   (Agrolux)

can pivot on the main frame. A parallel linkage system secures the bodies at the required setting. When a change in furrow width is required, the parallel linkage is adjusted either manually with a turnbuckle or hydraulically from the tractor seat. Either method provides very rapid adjustment from 300–500 mm (12–20 in), depending on the model of plough.

Controls and Adjustments

Hitching   Connect the plough to the three-point linkage in the correct order: the lower left-hand link, then the lower right-hand link and finally the top link. Remember 'Left—right—top'.

Adjust the top link length to the mid-position for transport. Both lift rods must be set at exactly the same length when using a reversible plough.

PLATE 9.6   *Slatted mouldboard.*   (Agrolux)

After attaching a conventional plough, set the right-hand lift rod in the mid-position.

When detaching the plough, remove the linkage in the reverse order.

PLATE 9.7   *Automatic reset mechanism: this plough has struck a block of concrete; the spring has allowed the leg and body to pivot backwards. Once the obstruction is passed, the body returns to the correct position.*

(Dowdeswell)

PLATE 9.8   *Variable furrow width linkage and ram on a reversible plough. Each furrow width is maintained at the same width by a parallel linkage system.*   (Kverneland)

Depth This is adjusted with the draft control lever and its stop. Some of the larger mounted and semi-mounted ploughs have an adjustable depth wheel.

Levelling Both lift rods must be set at the same length after attaching a reversible plough to a tractor. Screw adjusters in the headstock are used to ensure the plough bodies produce a level finish. The levelling box on the hydraulic linkage is used to maintain level furrows with a conventional plough.

Pitch Adjusted with the top link on all ploughs. Lengthen it if the heel iron rides clear of the furrow bottom and shorten it if the heel iron cuts a deep mark.

Too little pitch will make plough penetration difficult; too much will cause the shares to dig in and produce rough work.

Front furrow width Two adjusting bolts are usually provided to change the front furrow width on a reversible plough. Once set, however, it will rarely be necessary to use this adjustment. Incorrect front furrow width is often due to faulty setting of the levelling adjustment in the headstock.

A lever on the cross-shaft is used to alter the front furrow setting on conventional ploughs.

Incorrect front furrow width will cause very uneven work. It is quite easy to see the consequence of a wide or narrow front furrow with a high or low front furrow occurring in a regular pattern across the field. Check that the wheels are at the correct track setting for the plough. See the instruction book. It will be difficult to obtain the correct front furrow width if the track setting is wrong.

Setting the coulter

1. The disc coulter hub should be set over the point of the share.
2. The disc should be set 12 mm towards the unploughed side of the share. A stepped furrow wall indicates that the disc is set too far towards the unploughed land. A ragged furrow wall means that the disc is too far towards the ploughed land.
3. The height of the disc depends on ploughing depth. Set the bottom of the disc about 35 mm above the share. Set it higher for deep ploughing. In all cases, set the disc

coulter high enough to prevent the disc bearing housing from dragging on the soil, as this will cause undue wear.

A knife coulter should be adjusted so that the tip of the knife is in the same position as the bottom of a disc coulter.

Skim coulter setting This is set so that the skim coulter cuts a wedge, about 50 mm deep off the corner of the furrow slice about to be turned. The skimmer point should be below and behind the disc coulter hub. It should not be set too deep since excessive skimming can cause problems in gaining the required ploughing depth, especially on hard land.

Reversible Ploughs

Reversing mechanisms

At the end of each pass across the field, the plough is turned on its frame to bring the opposite set of bodies into work. The turnover mechanism may be either manual or hydraulic.

The tractor hydraulic system is used to operate an auxiliary service ram on the plough headstock. The double acting turnover ram is controlled by a remote valve unit in the tractor cab.

Some older models of reversible plough have a manual turnover mechanism operated by a hand lever. When the plough is lifted from work, its weight pushes forwards towards the headstock. The weight is used, by means of a special linkage, to rotate the plough when the driver pulls the hand lever. After the plough has been lifted from work and reversed, the turnover mechanism must be loaded before it can be used again. To do this, the plough is lowered and pulled forward in the ground.

Plough maintenance

- Grease all lubrication points daily during the ploughing season.
- Change all soil wearing parts when they are worn. This is particularly important with the shares because blunt shares make the plough hard to pull.
- Keep nuts and bolts tight at all times.
- At the end of the season, replace worn parts, coat bright surfaces with rust preventative solution and store the plough under cover.

PLATE 9.9 *Reversible plough headstock and turnover ram.* (Kverneland)

PLOUGHING SYSTEMS

A plough which is not properly set will need extra power to pull it through the ground. This wastes time and money. Always make sure that:

1. The plough is properly attached.
2. The tractor wheels are at the correct track setting for the plough.
3. The pitch adjustment is correct.
4. The mouldboards are parallel to each other. The mouldboard stays can be adjusted to rectify any fault in alignment.
5. The shares are in good condition.
6. The discs and skimmers are set deep enough to do their work but not too deep.
7. The plough is suitable for the tractor and does not have too many furrows for the power available to pull it.

Reversible Ploughing

For a tidy finish, in readiness for ploughing the headlands, the first task when starting to plough a field with a reversible plough is to mark headland furrows. The distance from the edge of the field to the headland furrow will depend on the number of furrows on the plough. For example, a three-furrow plough needs at least 9 m. An extra 2 m will make turning on the headlands easier and reduce wear and tear on the tractor. Larger ploughs will require a bigger headland.

It is best to plough the field longways: this reduces the number of headland turns needed. Start at one side of the field. On the first run turn some shallow furrows away from the hedge to disturb the vegetation. Then plough the moved soil back again and continue ploughing across the field. Finally, plough the two headlands to complete the field.

Conventional Ploughing

Systematic ploughing
Using a right-handed plough, mark a shallow furrow all round the field. This furrow should be about 8 m from the hedge for a three-furrow plough. Add 1 m for each extra furrow. Next, the field is marked off in sections with a shallow furrow. Each section—called a *land*—should be 30 m wide with an extra 10 m for each additional furrow. The first land should be only three-quarters of the full width.

A ridge is made by ploughing up and down on

PLATE 9.10 *Good reversible ploughing.*

either side of the marking furrow with the plough set for making an opening. When all the ridges have been made, the plough is reset for normal work and the field is ploughed by alternate casting and gathering.

Casting means working between two ridges with the tractor turning to the left on the headlands. Gathering means ploughing round and round a ridge with the tractor always turning right on the headlands. When the distance between the two furrows is narrow, the plough is reset to complete the ploughing by making a finish.

Systematic ploughing reduces idle running on the headland to a minimum. With the field completed, the headlands are ploughed. To help keep the field level, the headlands are ploughed towards the hedge one year and away from the hedge on the next.

Round and round ploughing

This method gives a surface with few furrows. Ploughing can be from the middle outwards or work can start at the edge and progress to the centre.

When ploughing outwards, the centre of the field must be found. To do this, a man walks round the outside of the field holding a cord, tied to the tractor. The driver keeps the cord taut, making a shallow marking furrow as he drives round the field. The helper next walks round the marking furrow, and the tractor driver makes another marking furrow nearer the centre of the field. This process continues until a small replica of the field shape is left in the centre. This is ploughed first, round a ridge, then the rest of the field is ploughed round and round.

With no help available and a reasonably shaped field, use this method. Measure the

shortest side of the field and divide it by two. Subtract 3 m from this measurement and then pace this distance from the hedge towards the middle, at intervals around the field. Walk into the field at right angles to the hedge. Mark each point with a stake and a small replica of the field shape will be produced in the middle. Plough the replica first, and then work round and round.

To keep a level field, it is best to work from the hedge, and then the centre, in alternate years. To plough from the hedge, first plough the headland and then make diagonal marking furrows. Continue ploughing round and round towards the middle but leave an unploughed strip on either side of the diagonal, marking furrows for turning. When the main part of the field is completed, the diagonal strips are ploughed.

SUGGESTED STUDENT ACTIVITIES

1. Learn the names of the main parts of a plough.
2. Locate the main adjustments on a plough and identify their purpose.
3. Hitch a plough to a tractor and make the basic settings for normal ploughing.
4. Always coat the bright parts of a plough with a rust preventative after use.
5. Study a specialist book on ploughing to find out how to improve your skill.

SAFETY CHECK

Never do any maintenance work to a raised plough without first supporting it with blocks or an axle stand.

Chapter 10

CULTIVATION MACHINERY

SUBSOILERS

Free movement of air in the soil and good drainage are essential for high crop yields. Tractors and machinery are mainly responsible for the soil compaction and soil pans on arable farms which reduce soil air movement and hinder drainage.

Subsoilers are used to break compacted soil, allowing the free passage of air and water. Mounted on the three point linkage, subsoilers have from one to five legs, fitted with shares which are pulled through the soil. Typical working depth is 400–500 mm. Trailed subsoilers with up to nine legs are made for track-layers and very large wheeled tractors. Power requirement is quite high: a typical three-leg

PLATE 10.1 *Power harrowing.* (Colchester Tillage)

89

PLATE 10.2   *A three-leg subsoiler with wings fitted to the shares. This design gives increased shattering effect on the subsoil.*   (Agrolux)

subsoiler, capable of working up to 550 mm, needs a 110 kW (150 hp) tractor. Each share has a replaceable point and it is usually possible to fit new shins—the front vertical cutting edge of the leg—when they are worn. For extra soil disturbance wing attachments can be fitted to the shares.

Greater shattering of the subsoil can be achieved by vibrating the subsoiler share below ground level. One type of subsoiler has a power take-off driven vibrator unit which agitates the shares as they pass through the soil. The agitation has a good shattering effect below the surface but does not heave the soil upwards to the same degree as a normal subsoiler. The vibrating subsoiler can have from two to thirteen legs depending upon model. The larger machines

can also have a three point linkage fitted to carry a second implement such as a power harrow or disc harrow. The power shaft is extended to provide the drive for power-driven implements.

Subsoiling is best carried out when the ground is dry and hard, for example after the pea or cereal harvest, to obtain the maximum shattering effect. It is usual to subsoil at right angles to the planned direction of ploughing. The spacing between passes of the subsoiler legs is from 1–2 m.

MOLE PLOUGHS

Heavy land requires draining to reduce its water content to a level satisfactory for efficient plant

PLATE 10.3 *Subsoiling.* (Agrolux)

Water from the moles seeps into the pipes and runs to the ditch.

The mole plough may either be trailed or mounted on the three point linkage. The heavier machines are usually pulled by a crawler, but the lighter mounted models make mole ploughing possible with a powerful wheeled tractor. The machine consists of a very strong frame which slides along the ground when the mole plough is in work. A heavy leg, similar to a subsoiler leg, is attached to the frame. A circular section share with a larger diameter expander on a flexible link is bolted to the leg. The share forms a tunnel, about 75 mm diameter in the soil, and the expander presses the soil outwards to form a long-lasting drainage channel.

CHISEL PLOUGHS

The chisel plough has a heavy duty frame with a number of tines bolted to it. Each tine has a replaceable point. Sometimes a shear bolt is incorporated at the foot of the tine. When the tine hits an underground obstruction, the shear bolt breaks, allowing the point to swing backwards. Chisel ploughs are tractor mounted, and working depth is hydraulically controlled. Some

growth. Heavy soils usually have a system of permanent drains using either clay or perforated plastic pipes which discharge into a ditch. A mole plough is used to make mole drains in the soil at a depth of up to about 950 mm. These are drawn through the soil at an angle to the pipe drains.

PLATE 10.4 *Tractor-mounted mole plough. The expander can be seen on the ground behind the point. Much larger wheeled mole ploughs are used with heavy tracklaying tractors.* (Agrolux)

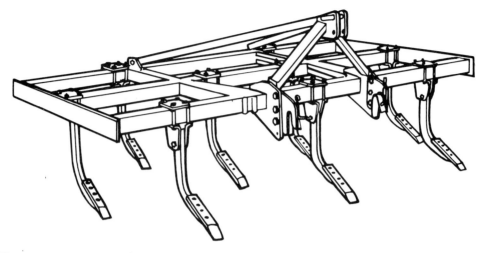

FIGURE 10.1   Chisel plough.   (*Agrolux*)

of the larger chisel ploughs have wheels on the frame to regulate working depth.

Chisel ploughs are made in widths ranging from 2–5 m. The tines are arranged in two or three rows and the number used depends on the type of work, power available and soil condition. Rigid tines are normally used but spring-loaded or heavy duty flexible tines are available. Power requirement for chisel ploughing varies from 45–120 kW (60–160 hp).

Uses

Chisel ploughing is considered by some farmers to be an alternative to mouldboard ploughing, especially in arable areas where there is little surface rubbish to bury. The reduction in stubble burning, due to resistance to this practice by the public, has seen a decline in the use of chisel ploughs.

Heavy cultivating, especially to break stubbles after harvest, is often done with a chisel plough. It can also be used as a subsoiler if all but three of the tines are removed.

Cultivators

A cultivator has a frame with a number of tines for breaking and stirring the soil. It is usually tractor mounted and working depth is controlled hydraulically. Large mounted cultivators often have depth control wheels. As with all cultiva-

tion machinery, there is a wide range of working widths, from 2–8 m, to suit all sizes of tractor.

Types of tine

Rigid tines   Used for heavier work, the tines are staggered across the frame to allow free passage of soil and reduce blockages.

Spring-loaded tines   These are non-flexible tines, held in their working position by heavy springs. The tines lift a little on hitting a large clod, the spring pressure pulling the tine down again, and this action helps to shatter the soil.

Spring tines   These are flexible, square or flat sectioned, sometimes with a coil at the top, which vibrate in the soil. The vibrating movement of the tines will give fast seedbed preparation in most conditions.

Types of share

The reversible share is used for seedbed and general work. Duckfoot and broadshares are mainly used for stubble breaking and cleaning. When worn, the shares must be replaced.

The uses of cultivators

1. Seedbed preparation, especially after ploughing.
2. Stubble cleaning and breaking.
3. General weed control.

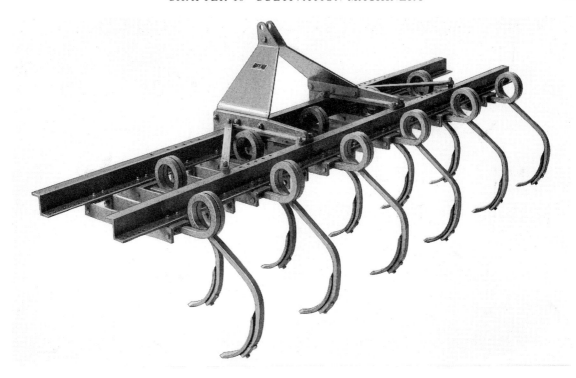

PLATE 10.5   *Spring tined cultivator.*   (Agrolux)

4. Cultivating between potatoes and other rowcrops. The tines are grouped to pass between the rows without disturbing the crop.

One-pass Cultivators

Designed for minimal cultivation techniques, these machines have three sets of tines which work progressively deeper from front to rear. The front blades break up the topsoil. The centre section has a staggered row of heavy duty cultivator tines working at a medium depth. They are followed by subsoiler legs attached to the back of the cultivator.

A typical 3.6 m model requires a tractor of at least 135 kW (180 hp).

POWER-DRIVEN CULTIVATORS AND HARROWS

Power take-off-driven cultivation machines include power harrows and rotary cultivators.

Power harrows have two or more rows of reciprocating tines, others have tines which rotate in the soil. All power-driven cultivation machinery is costly and needs regular maintenance to ensure a long working life. An advantage of power harrows is their ability to prepare a seedbed without bringing up unweathered soil from below the surface. This gives a clod-free tilth.

Rotary Cultivators

These may be mounted or trailed and have a power take-off-driven shaft with 'L' shaped blades which cultivate the soil. The drive arrangement is through a gearbox and chain drive unit. A drive disengagement lever is sometimes included in the gearbox, allowing the power shaft to turn with the blades stationary. An overload slip clutch is also built into the drive.

The rotor and blades run at speeds ranging from 90–240 rpm with the power shaft turning at 540 rpm. The blades throw the soil against a

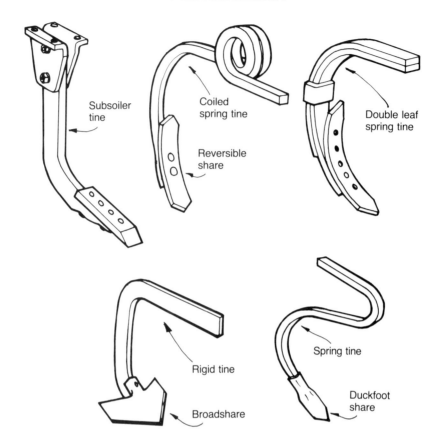

FIGURE 10.2   Types of tine and share.

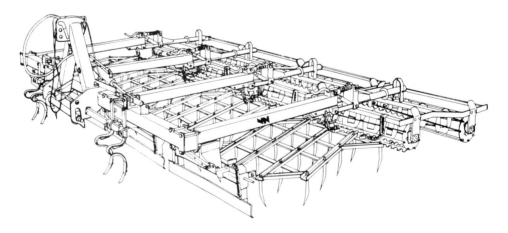

FIGURE 10.3   Combination harrow with levelling boards, harrows and two crumbler rolls. Wheel mark eliminating tines are fixed at the front of the machine.   (*Vogel and Noot*)

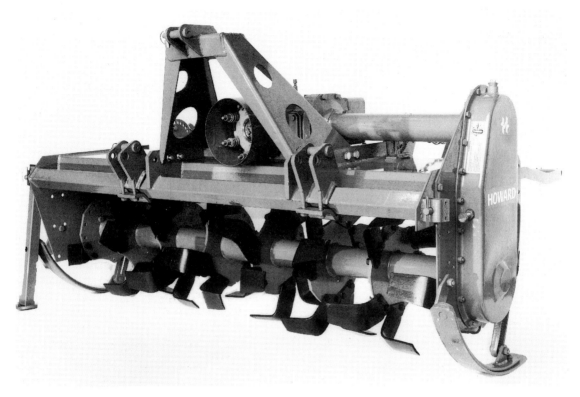

PLATE 10.6   *Mounted rotary cultivator. The overload slip clutch springs can be seen below the top link pin. Drive from the gearbox is transmitted to the rotor by a heavy chain in the chain case. Note the depth control skids at each side of the machine.*   (Howard Farmhand)

hinged flap at the rear of the machine; this helps to shatter the clods. Rotary cultivators are made in working widths ranging from 1 m to more than 5 m; the wider versions are trailed and require tractor power in the region of 110 kW (150 hp).

Adjustments

Tilth   Three adjustments can be made to vary the type of tilth produced:

1. The rotor speed can be varied by choosing a different gear ratio in the cultivator gearbox. A fast rotor speed gives a fine tilth.
2. The tractor forward speed, changed through the gearbox with the power shaft speed maintained at 540 rpm. A low forward gear will give a fine tilth.
3. The hinged flap behind the machine gives a fine tilth when lowered. The tilth will be very coarse with the shield raised.

Working depth   Adjusted with a wheel. Depth of trailed models is controlled with an external ram on the machine. A skid is sometimes fitted to prevent the cultivator going too deep, especially if the depth wheel drops into a low place.

Uses

1. Stubble cleaning.
2. Seedbed work, especially for root crops. A spiked rotor can be used for seedbed making.
3. Cutting up weeds and crop residues.
4. Land reclamation work and general tidying around headlands.

Power Harrows

These have become a very important tool in the production of seedbeds. There are two types,

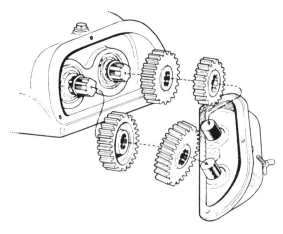

FIGURE 10.4   Rotary cultivator gearbox used to change
the rotor speed.

reciprocating tine and rotary. Most farmers now
use the rotary type which is available from a
number of manufacturers and importers.

Rotary Harrows

Rotary power harrows have vertical rotating
tines fixed to a series of rotor heads across the
full width of the machine. The drive is from the
power take-off through a gearbox and system
of gearing to the tine rotors. There is a choice
of rotor speeds, with a range of 200–450 rpm
available depending on model. For most work,
a rotor speed of 250–350 rpm is recommended.

Working widths vary from 1.5–8 m. The
smaller power harrows are mainly for horticultural use and have a lightweight design. Some
of the larger models are moved from one field
to another on a transporter, while others can
be hydraulically folded for transport. Power
requirement is high, ranging from about 38 kW
(50 hp) for the smaller power harrows up to a
massive 185 kW (275 hp) for the largest models.

A basic model of rotary power harrow has
a lightweight frame to support the gearbox and
rotors. A levelling bar and simple crumbler roll
are attached to the frame at the rear. A heavy
rod-type crumbler roll or a coil crumbler as well
as the levelling bar is fitted to the larger harrows.

Some power harrows have an auxiliary three
point linkage and power shaft extension allowing a drill or planter to be direct coupled to the
harrow. The drill and harrow are both lifted
when turning on the headland. An alternative
method of attaching a drill behind a power
harrow is a bridge link. This is used to couple a
trailed machine behind the harrow yet still allow
the driver to lift the harrow from work with the
tractor hydraulics.

PLATE 10.7   *Rotary power harrow.*   (Colchester Tillage)

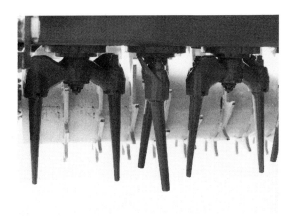

PLATE 10.8 *Rotary power harrow tines are timed to prevent them hitting each other.* (Krone)

PLATE 10.9 *A power harrow with linkage for attaching a drill to give a combination of harrowing and drilling in one pass. A clod breaking roller is fitted behind the power harrow.* (Colchester Tillage)

Yet another option is to fit the power harrow on the front linkage and the drill on the rear linkage. Other machines, such as a sprayer or fertiliser spreader, can also be fitted to the rear linkage, considerably increasing the work rate.

Adjustments
The fineness of tilth produced with a rotary power harrow depends upon the speed of the tine rotors and the forward speed of the tractor. A fine tilth is achieved with a low gear and high rotor speed.

Working depth, up to a maximum of 250 mm, is adjusted by raising or lowering the crumbler roll.

Reciprocating Bar Harrows

This type has two or four reciprocating tine bars driven by the power take-off through a gearbox and eccentric unit. They are three point linkage mounted and depth control is by the rear crumbler roll. Working widths range from 1.5–4.5 m. The tines can work down to a depth of about 200 mm if required.

Power harrow maintenance
Frequent lubrication is important. Gearbox oil levels need checking. Make sure all nuts and bolts are tight, paying special attention to the tine retaining bolts. The tines need to be replaced when worn.

DISC HARROWS

Disc harrows cut and consolidate the soil. Two or more sets of saucer-shaped discs are fixed to a frame which may be mounted or semi-mounted. Some heavy duty discs are trailed and have hydraulically operated transport wheels. Disc diameter varies from 300–750 mm. Each set of discs is supported by a pair of plain steel or ball bearings. Disc harrow working widths vary from 1.5–6 m.

Adjustments

Disc angle A hand lever or hydraulic ram on the harrow is used to alter the cutting angle of the discs. When set at the widest angle to the direction of travel, the soil movement will be greatest. When straight, the discs will not move the soil very much and will have a consolidating effect.

Working depth is controlled with the tractor hydraulics or ram operated depth wheels. It is often necessary to add extra weight to obtain the required working depth. Some disc harrows have weight trays on the frame for this purpose.

Scrapers should be set to keep the discs clean.

FIGURE 10.5   Reciprocating power harrow with crumbler roll.

Uses

1. Seedbed work, especially after ploughing in grass, where a tined implement would pull the turf to the surface.
2. Cutting up crop remains like kale stalks.
3. Some farmers use heavy duty disc harrows, sometimes with cutaway discs for stubble breaking or even as an alternative to ploughing.

Maintenance

Regular greasing, and checking the tightness of nuts and bolts are the main needs of disc harrows. The discs must not be allowed to work loose on their shaft.

Harrows

Spike Tooth Harrows

These consist of simple frames with tines bolted to them where the frame members cross. There is a variety of harrow sizes and weights. Light harrows are used for final seedbed work and covering seeds. Medium and heavy harrows are used for breaking down ploughed land. Four or more harrows are attached to a trailed harrow pole or a frame mounted on the three point linkage.

Harrow tines must be kept tight. If allowed to work loose, they will wear the holes oversize, making it impossible to keep the tines tight.

Dutch Harrow

Sometimes known as a float. This is a useful implement for sugar beet seedbed work. The metal or wooden frame has heavy tine bars which are pulled almost at right angles to the direction of travel. The tines loosen the soil and the heavy bars level the surface. Sometimes a crumbler roll is attached to the rear to give further treatment to the soil.

Chain Harrow

Similar in construction to a piece of chain link fencing. The links may be plain or spiked. Plain link chain harrows can be used to spread droppings and molehills on pastures or cover seeds. The spiked type is used to aerate grassland.

Rolls

The two main types of roll are the flat roll and the Cambridge or rib roll. The Cambridge roll has a number of cast iron rings on an axle which leave a corrugated surface. The flat roll may be a number of cast iron rings with a flat surface, a lightweight steel cylinder or a much heavier steel cylinder which can be filled with water to make a very heavy roll for grassland.

Rolls for smaller tractors are usually in sets of three—one wide and two narrow—giving a

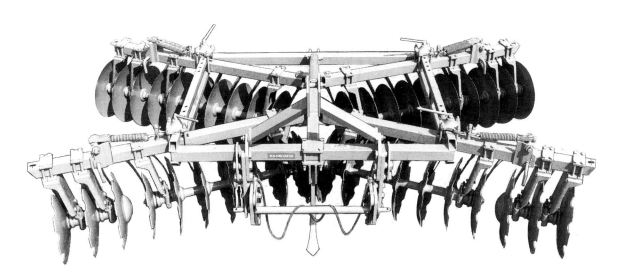

PLATE 10.10   *Mounted tandem disc harrows.*   (Agrolux)

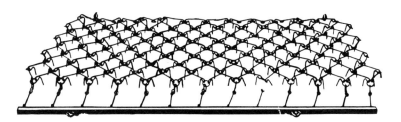

FIGURE. 10.6   Chain harrow.

working width of up to 8 m. For transport, the rolls are hitched one behind the other.

Much wider rolls for higher powered tractors are also available. These are usually in sets of five and are folded and unfolded by means of hydraulic rams. The roll sections are carried on a heavy frame which has pneumatic wheels used for transporting them on the road. This type of roll can have working widths of up to 12 m.

The axle bearings need regular greasing when in use. The roll sections must be kept tight on their axles. The sections rub against one another and wear quite quickly. Loose sections may break, especially when pulled along the road. Large washers are made to fit on the roll axle, keeping the sections tight.

Uses
Cambridge rolls:

- Breaking clods.
- Firming the soil.
- Pushing stones into the soil to reduce the risk of them damaging harvest machinery.
- Leaving a finely ridged surface for broadcasting grass seed.

PLATE 10.11   *Five gang roll hydraulically folded for transport.*   (Lely)

PLATE 10.12   *Furrow press with crumbler roll. The walking stick shape drawbars are automatically picked up by a bracket on the plough when it starts a new furrow. The linkage is arranged so that the correct drawbar is positioned for the next bout of ploughing.*   (Colchester Tillage)

Flat rolls:

- Breaking clods.
- Providing a smooth, firm surface for drilling.
- Pushing in stones on pastures (especially with the very heavy or water-filled type).

FURROW PRESSES

It is important, when ploughing in dry conditions, especially in the autumn, to retain soil moisture. Rolling as soon as possible after ploughing is one solution to this problem. The furrow press offers an alternative. It is a narrow version of a Cambridge roll with one or two sets of thin, widely spaced rings. The press is pulled from a special drawbar bolted to the plough. The working width can vary from approximately 1–3 m, with models to suit three- to seven-furrow ploughs. Some furrow presses have a three point linkage for transport purposes. An extra crumbler roll is sometimes connected behind the main press.

In operation, the furrow press is automatically detached when the plough is lifted at the headland. The furrow press has a double pick-up arm which provides a method of attaching it to a reversible plough in both directions of travel. After turning on the headland, the plough is lowered into work and the towing hook on the plough engages with the furrow press pick-up arm. Care is needed when lowering the plough to ensure that the press is coupled successfully.

TOOLBARS

These can be mounted at the front, underneath or behind a tractor. A wide range of implements, including ridging bodies, cultivator tines, hoe blades, etc. can be fitted to suit various rowcrop widths. The tractor wheels must also be set to suit the row widths.

Inter-row hoes may be fitted on a front-, mid- or rear-mounted toolbar. Front hoes are very sensitive to slight movements of the steering wheel but give good visibility of the row. Mid-mounted hoes are less sensitive to steering changes but visibility of work is not good. Traditionally, rear-mounted hoes had a second operator to steer the blades close to the rows, but this practice has almost ceased. Rear-mounted hoes are now designed for one-man operation.

'L' shaped or side hoe blades are used to cut weeds close to the rows and 'A' hoes remove weeds between the rows. Some inter-row hoes have plant protection shields to prevent soil being thrown over the plants when they are small.

Rowcrop work with potatoes using ridging bodies, cultivator tines or weeders is another important use of tractor toolbars.

Self-propelled toolbars, which consist of a toolframe with an engine and transmission system, are used by some farmers who grow rowcrops on a large scale.

PLATE 10.13   *Inter-row hoe. The plants are protected by the shields.*   (Stanhay Webb)

SUGGESTED STUDENT ACTIVITIES

1. Make a list of the cultivation machinery on a farm you know. Find out what each implement is used for.
2. Look at seedbed work in progress and study the effect that each implement used has on the soil.
3. Use the different adjustments on a power harrow or a rotary cultivator and observe their effect.

4. Find out how the hoes can be set on a toolbar to suit different row widths.
5. Study the operation of the linkage mechanism used for towing a furrow press behind a reversible plough.

SAFETY CHECK

Take great care when working up and down slopes. Use a gear which will give engine braking when going downhill, have sufficient front weight and never drive downhill with the gearbox in neutral.

Chapter 11

DRILLS

Grain Drills

There are two basic types of grain drill. One has a mechanical feed mechanism which supplies grain at a controlled rate to the drill coulters. Pneumatic drills, of more recent introduction, have a mechanical seed metering unit with an airflow system which distributes the grain to the coulters.

The more common type of drill, which has been in use in one form or another for more than a century, has a box or hopper, either three point linkage mounted with small pneumatic drive wheels or carried on large diameter wheels. The hopper contains a mechanical feed mechanism which delivers grain to the seed tubes. These supply the grain to the coulters which make a shallow furrow in the soil. The grain is covered

PLATE 11.1 *A pneumatic drill in action. Note the furrow left by the marker disc to ensure an accurate join between bouts.* (Massey-Ferguson)

by tines attached to the coulters or by a set of following harrows attached to the drill or another tractor.

Grain Drill Feed Mechanisms

External force feed A shaft beneath the hopper carries a set of revolving fluted rollers, one for each coulter. The flutings carry grain from the hopper to the seed tubes at the required seed rate. This mechanism is only suitable for cereal crops. Some versions have an adaptor plate close to the fluted rollers. The clearance between plate and roller can be varied, making it possible to drill a wider range of seeds. The seed rate is altered by:

1. Changing the length of fluted roller in contact with the seed. A blanking plate, controlled with a lever, is used to adjust the amount of fluting which meters the flow of grain. A graduated scale against the lever indicates the seed rate setting.
2. The speed of the fluted rollers can be changed by means of a gearbox. Maximum

seed rate is achieved with the highest roller speed and the full length of the fluted rollers in contact with the grain.

Studded roller feed This is another form of external force feed. The fluted rollers are replaced with studded or pegged rolls. Seed is carried from the hopper to the seed tubes by the studs as the mechanism turns on its shaft.

Seed rate is controlled by:

1. The speed of the studded rolls.
2. The position of an adjustable flap which controls seed flow from the hopper to the feed mechanism.
3. The type of studded roll. With some drills, rolls with different sizes and numbers of studs can be used.

The studded roller feed is suitable for drilling most types of seed.

Internal force feed A number of discs, one for each row drilled, turn on a shaft at the bottom of the hopper. Each disc has a wide rim on one side and a narrow rim on the other. The wide rim is

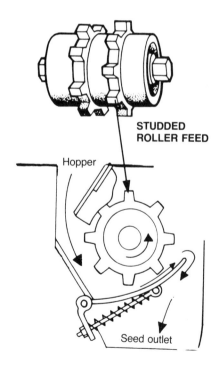

STUDDED
ROLLER FEED

Hopper

Seed outlet

EXTERNAL
FORCE
FEED

FIGURE 11.1 Studded roller feed and external force feed.

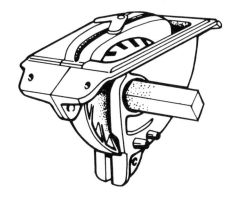

FIGURE 11.2   Internal force feed.

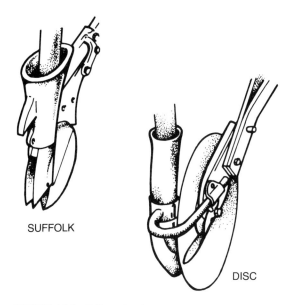

SUFFOLK

DISC

FIGURE 11.3   Drill coulter types.

mainly for drilling peas and oats and the narrow one for wheat and barley. A cover is placed over the unwanted side of the feed disc. Serrations on the inside of the rim pick up seed as the discs turn in the hopper and carry it to the seed tubes. The seed tubes connect the outlets under the hopper to the coulters.

The amount of seed sown per hectare (the seed rate) is adjusted by:

1. Changing the speed of the feed discs. Most drills have a gearbox to alter the speed of the discs. Sometimes different gears are fitted to the drive shaft for this purpose.

2. For high seed rates with wheat or barley, the wide rim of the feed disc is used.

Grain drill feed mechanisms are driven by the drill wheels and a method of disengaging the drive is always provided.

Coulters

'*Suffolk*'   This has a cast-iron shoe which cuts a narrow furrow for the seed. Its shape helps to keep a straight and even depth furrow. Suffolk coulters do not block easily and are not affected by stony land. There are no moving parts to lubricate but soil contact wears the bottom of the coulters rather quickly. Penetration can be difficult in hard ground. Suffolk coulters are suitable for sowing both root and cereal crops.

Disc coulters are saucer shaped and cut a rather irregular depth furrow. Their cutting action helps with penetration and they work well in most soils. Stones can stick between the disc and its support causing uneven disc wear. Disc coulter bearings need frequent lubrication. They are more expensive than the Suffolk type but will last longer if properly maintained. Disc coulters are widely used for drilling cereal crops but are not suitable for root crops.

Hoe coulters   Their main use is for minimal cultivation techniques where chisel ploughs or heavy disc harrows are used instead of mould-board ploughs. Narrow, replaceable points are fitted to spring tines of the type used for spring tine cultivators. The tines stir the soil and make a shallow furrow for the seed. The seed tubes from the feed mechanism are attached to the back of the coulter tines.

Sowing depth and row width

An even sowing depth is important. The coulters are carried on arms which pivot from the front of the drill. Each coulter has a pressure spring to force it into the ground. The spring pressure can be adjusted to suit soil conditions and give the required sowing depth.

Some drills have a tractor-operated hydraulic ram to set and maintain the required depth.

The coulters are lifted from work by a mechanical lift clutch operated by the drill wheels, a hand lever or a hydraulic ram.

Most drills have a row width of 180 mm. Others have row spacings of 90 or 120 mm. Some drills have adjustable row widths to suit a range of crops. The coulters are spaced across the drill to give the required row width. Working widths of grain drills vary from 2.5–6 m.

Seed tubes

Rubber or plastic tubes are in common use. They will bend easily when the coulters are lifted and are not corroded by fertiliser.

Spirally wound metal seed tubes are flexible but will rust and are easily damaged.

Telescopic tubes, made of metal or plastic, allow the coulters to move up and down on uneven surfaces. There is no problem with bent or misshapen tubes if the drill is stored with the coulters raised.

When storing drills, the seed tubes must hang straight or they will be bent or kinked when the drill is prepared for the next season. The best solution to this problem is to store the drill with the tubes removed from the coulters.

Pneumatic Grain Drills

This type of drill is capable of drilling a wide range of seeds at relatively high forward speeds.

Both trailed and mounted versions are available with working widths of 2.5–8 m in common use.

The pneumatic drill seed metering mechanism is driven by the land wheels. Seed from the hopper is metered at a pre-set rate to a venturi where it is introduced to an airflow. The power take-off driven fan blows the grain to the distributor head where it is evenly supplied to the seed tubes. The airflow carries the seed along the tubes to the coulters. Once in the soil, the seed coverers close the furrows. Some drills have a hinged flap which prevents soil blocking the coulters if the drill is reversed while in work.

Seed rate is altered by adjusting the metering unit with a lever or handle. The air blast fan must run at the recommended speed for efficient operation. Sowing depth is controlled hydraulically.

The wider models of pneumatic drill may have two hoppers with their own metering units which deliver grain to the air distribution system. The pneumatic drill is often used in conjunction with a power harrow. The drill is attached to the power harrow rear linkage, and an extension to the power take-off shaft drives the fan. When lifted from work the drill moves upwards and forwards over the harrow to improve stability.

PLATE 11.2   *Front-mounted grain hopper supplying seed to a pneumatic drill on the back of the tractor.*   (Ferrag)

COMBINE DRILLS

Some farmers prefer to drill both grain and fertiliser at the same time with a combine drill. The grain side will use one of the feed mechanisms described above, but the hopper is much larger and is in two parts, one for seed, the other for fertiliser.

Fertiliser Feed Mechanisms

Roller feed A mechanism very similar to that used on grain drills is the usual type of fertiliser feed mechanism on combine drills. The various parts are made of plastic, as is often the case for the grain feed, to overcome corrosion problems. Application rate is altered by:

1. Changing the speed of the feed rolls with a gearbox lever.
2. The size of the hopper outlet can be changed with lever-operated adjustable shutters.

Star wheel Now outdated, this mechanism consists of a series of star-shaped wheels rotating in the hopper bottom over outlets which supply the fertiliser to tubes connected to the coulters. Application rate is controlled by:

1. The speed of the star wheels through a gearbox lever.
2. The size of the outlet in the hopper bottom can be changed with lever-operated adjustable shutters.

Some combine drills use one tube to deliver both seed and fertiliser to the coulters; others have separate tubes for each material.

It is very important to clean combine drills thoroughly after use as the corrosive action of fertiliser will damage them.

Direct Drills

The technique of direct drilling means that the seed is placed directly into undisturbed soil such as a stubble field. Direct drilling has lost favour

PLATE 11.3 *A combine drill sowing both grain and fertiliser at the same time. Spring tines cover the seed and scratch out wheel marks. The drill transport wheel can be seen in the raised position.* (Massey-Ferguson)

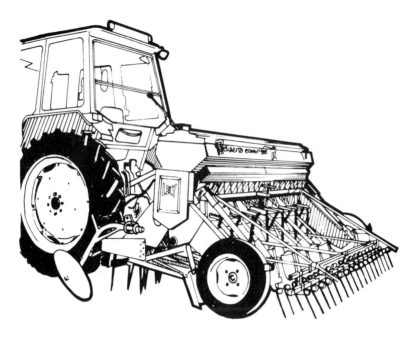

FIGURE 11.4 Cultivator drill.

with most farmers. The direct drill has very heavy coulters which cut grooves in the soil. The seed tubes place the grain or other seed into these slots which is then buried with some form of coverer.

Another type of direct drill has very heavy spring tines with narrow shares. The seed tubes drop the grain into the soil immediately behind the shares.

Direct drills are sturdily built to withstand the high rate of wear and tear which must be expected when working on hard land. For this reason they also tend to be rather expensive.

Weed control can be a problem with direct drilling. The surface must be sprayed a few days before drilling to kill off the vegetation. This method of drilling can also be used after burning off both straw and stubble, another technique which is no longer popular, especially with the general public.

Cultivator Drills

Normally used on ploughed land, this type of drill consists of a tined cultivator with a drill mounted above the tines and a small following harrow to cover the seed. The seed tubes are attached to the tines. A more expensive version is a power harrow and drill combination with either a conventional or pneumatic seeding mechanism.

Seeder Units

Mainly used for root and vegetable seeds, seeder units have a feed mechanism which places seed in the soil at a variety of spacings to suit a wide range of crops.

A number of seeder units, usually five or six, but sometimes up to twelve, are attached to a tractor-mounted toolbar. Each unit has a frame supported by two wheels. The frame has a coulter, a covering device and a seed hopper, complete with feed mechanism attached to it. The rear wheel firms the soil after the seed is covered. The feed mechanism may be driven by:

1. The front wheel of the seeder unit, with each unit individually driven.
2. Master wheel drive where the main land wheels, supporting the toolbar, drive all the seed mechanisms through a system of chains or belts and a gearbox.

PLATE 11.4 *Precision seeders on the toolbar. Drive is from the land wheels to the gearboxes at the top of the bar.* (Stanhay Webb)

Seeder Unit Feed Mechanisms

Belt feed A small, endless rubber belt, which has holes punched in it, carries seed from the hopper to the outlet point. A repeller wheel, turning in the opposite direction to the belt, helps to ensure that only one seed is carried in each hole. The belt is stretched slightly at the outlet point and the seeds drop through the holes into a shallow furrow made by the coulter.

Seed rate and spacing are altered by:

1. Changing the seed belts. A range of belts is available with holes punched at different intervals. Belts with different size holes are used to suit a variety of seeds.
2. Belt speed can be changed on master wheel seeder units. A gearbox drive provides a range of belt speeds. Only one set of belts may be needed to give a full choice of seed rates and spacings for a particular crop. The choke controls seed flow from the hopper.

Cell wheel feed Holes in the rim of the cell wheel collect single seeds from the hopper. A repeller wheel, turning in the opposite direction to the cell wheel, helps to ensure that only one seed enters each hole. The seed is carried by the cell wheel to the outlet point where an ejector plate prises the seed from the holes and they fall to the ground.

Seed rate and spacing are changed in the same way as for belt feed. The cell wheels are changed as required. Different size holes cater for different types of seeds.

A selection of seeding belts with different numbers and
sizes of hole

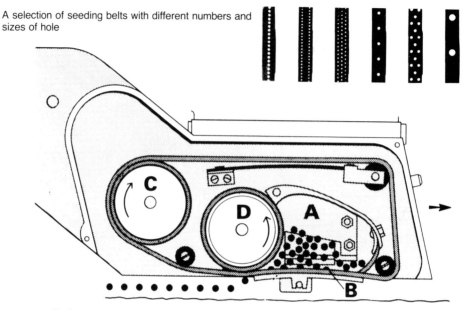

A – Choke
B – Seed belt
C – Drive wheel
D – Repeller wheel

FIGURE 11.5   Belt feed mechanism.   (*Stanhay Webb*)

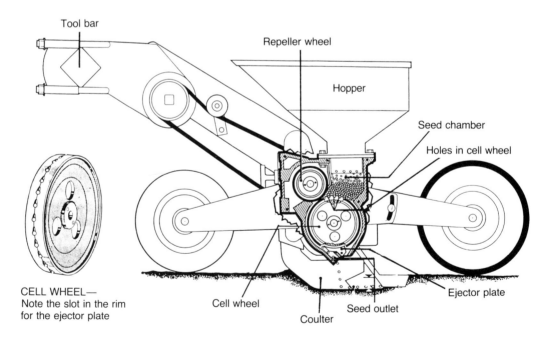

CELL WHEEL—
Note the slot in the rim
for the ejector plate

FIGURE 11.6   Cell wheel feed mechanism.   (*Stanhay Webb*)

Warning systems Seeder units usually have a set of warning lights to tell the driver if any of the feed mechanisms are not working. A light is also provided to show the driver when the seed hoppers need refilling.

Using Drills

Markers
These are used to help the driver join his work accurately. Sometimes the drill wheelmarks can be used as a guide. Markers, one each side of the drill, have a pointed tine on a bar, which is fixed to the drill. It makes a mark to show the driver where to drive on the next run. The position of the tine can be adjusted. To set drill markers, follow this method:

1. Measure the distance from the centre of the tractor front wheel to the outside coulter (distance A).
2. Measure the row width (distance B).
3. Add distance A to distance B. The marker point should be set at this distance from the outside coulter.

Markers are operated manually on older drills with a hand lever either on the back of the drill or within reach from the tractor seat. Markers on modern drills are usually operated with a hydraulic ram controlled from the tractor cab.

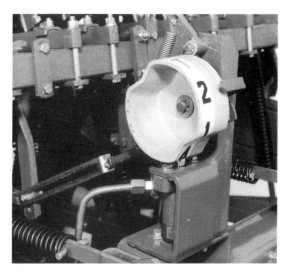

PLATE 11.5 *Tramlining equipment on a grain drill. The cam unit moves one step each time a bout is completed. The mechanism is operated by a linkage from the coulter lifting gear. An over-ride mechanism is used if the coulters have to be lifted to clear a blockage while drilling.* (Massey-Ferguson)

Tramlines
Tramlines are the unsown rows which provide wheelings for spraying and fertiliser spreading after the crop has emerged. This is a technique favoured by many farmers since it avoids the tractor wheels damaging the growing crop and ensures accurate joining of bouts when using sprayers or fertiliser spreaders.

There are two basic types of tramline equipment, manual and automatic. When first introduced, the controls were manual and relied on the tractor driver to remember to turn off the drill feed to the two rows which form the tramlines at the proper time. Automatic tramlining depends on an electronic control box in the cab. One model gives the driver a choice of one to six bouts per tramline sequence. The tramline control unit is operated through the automatic marker changeover mechanism. It receives signals from the markers and, after the pre-set number of bouts is completed, the tramliner solenoid shuts off the feed mechanism to the two rows which form the tramline. An over-ride switch is provided for use if the markers have to be raised during the bout to clear blocked coulters.

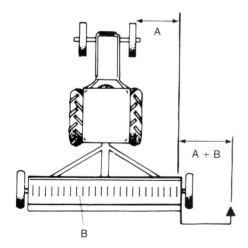

FIGURE 11.7 How to set drill markers.

Using drills

Most grain drills are moved from field to field by towing the drill from the normal working position. The very wide drills, 3- and 4-m models for example, are transported with a drill trailer or a towing kit on the drill. Hydraulically lowered transport wheels are fitted to one side of the drill with a removable drawbar at the other. The wheels are raised and usually removed together with the towbar in preparation for drilling.

A tractor and trailer will be required to transport the grain and fertiliser along the headland as drilling proceeds. It is usual to drill three or four bouts around the headland before drilling the main part of the field. This gets the headlands drilled before the soil is compacted when turning at the end of each bout. It also provides a mark for lifting and lowering the drill coulters into or out of work.

When using a cultivator drill or a drill/power harrow combination, some farmers prefer to drill the headlands last.

Checking the Sowing Rate

The amount of seed sown per hectare (acre) can be checked or calibrated in the field or barn.

A field check is made by setting the drill at the required rate and then drilling a measured area with a known quantity of seed. For example, to check a seed rate of 125 kg/ha (1 cwt/acre), put 125 kg of grain in the hopper. This should be used up just as the hectare is completed. A drill can be calibrated in the barn by turning the drill wheel the same number of times as it would turn to drill a certain area. The feed mechanism drive is engaged and the grain is collected and weighed at the end of the check.

To find the number of turns of a drill wheel to cover an area of the field, use this formula:

$$\text{Number of wheel turns for } 1/25 \text{ hectare} = \frac{400}{\underset{\text{(m)}}{\text{Sowing width}} \times \underset{\text{(m)}}{\text{Wheel circumference}}}$$

$$\text{Number of wheel turns for } 1/10 \text{ acre} = \frac{484 \times 9}{\underset{\text{(ft)}}{\text{Sowing width}} \times \underset{\text{(ft)}}{\text{Wheel circumference}}}$$

The drill is jacked up and supported with blocks, so that the driving wheel can be turned by hand. Partly fill the hopper with grain and set the drill to give the required seed rate. Place sacks or sheets under the coulters, engage the

PLATE 11.6   *Grain drill in transport position. The drawbar has been raised to a vertical position and the transport wheels lowered by the auxiliary ram above the wheel.*   (Massey-Ferguson)

drive and turn the wheel for the number of times calculated. Collect and weigh the grain. The seed rate per hectare is found by multiplying the grain weight by 25. (Multiply the grain weight by 10 if the second formula is used.)

PLATE 11.7 *Calibrating a grain drill. The crank handle* (right) *is used to turn the feed rollers a set number of times.* (Massey-Ferguson)

Where the seed rate is found to be inaccurate, reset the drill and repeat the process until the correct setting is found. Make a permanent record of the correct settings.

Very wide drills may have half the feed mechanisms driven by each land wheel. In such cases, both wheels must be turned to complete the calibration.

The calibration method described above is for drills with large diameter wheels. Many drills, both mounted models with small wheels and trailed drills, can be calibrated in a simpler way, using the calibration cranking handle which is provided: turn the feed mechanism a stated number of times after setting the adjustment to give the required seed rate. Calibrating trays on the drill collect the grain which is then weighed to find the actual seed rate.

Maintenance and storage of drills

- Lubricate every day when in use.
- Check that the coulters, seed tubes and feed mechanisms are in good condition.
- Keep all belt and chain drives at the correct tension. The fertiliser hopper of a combine drill should be emptied at the end of the day.

PLATE 11.8 *Calibrating a grain drill. Withdrawing the tray used to collect grain from the feed rolls when they are turned by the crank handle. The stub end for the handle can be seen protruding from the drive guard.* (Massey-Ferguson)

This will reduce damage caused by the corrosive effect of fertiliser.

Drills must be cleaned thoroughly before storage. All traces of fertiliser must be removed from combine drills. Coat bright surfaces with rust preventative and lubricate all oil and grease points. Make sure that the seed tubes hang straight or remove them completely. Store the drill under cover.

SUGGESTED STUDENT ACTIVITIES

1. Trace the line of drive from the land wheels to the feed mechanism of a corn drill.
2. Find out how the grain and fertiliser application rates are changed on a combine drill.
3. Study cell wheel and belt feed seeder mechanisms and note the methods used to alter seed rate.
4. Calculate how to set the markers on a grain drill.
5. With the aid of an instruction book, calibrate a grain drill.

SAFETY CHECK

Remember to use a securing clip in the drawbar pin when towing a grain drill or other trailed implement.

Chapter 12

MANURE AND FERTILISER DISTRIBUTION MACHINERY

FARMYARD MANURE SPREADERS

A manure spreader is basically a trailer with a moving floor conveyor and a combined shredding and spreading mechanism at the rear which distributes farmyard manure on to the land. Manure spreaders are power take-off driven. A few land wheel drive machines are still in use.

Another type of manure spreader has an open sided cylinder shaped trailer body with a rotor shaft running lengthways through the centre. Flails, fixed to the rotor shaft by heavy chains, throw the manure from the side of the spreader to the ground. Rotary spreaders are also power

take-off driven, and can be used for both liquid and solid materials.

Rear Delivery Spreaders

The spreader floor has an endless chain and slat conveyor which moves the manure to the rear of the machine. Here, high speed shredders tease out the material and spread it, depending on the model, up to a width of about 12 m.

The spreading mechanism varies from make to make. Some machines have a single rotor which both shreds and spreads, others have as many as three shredders. Yet another system uses

PLATE 12.1   *A spinning disc broadcaster top dressing pasture.*   (Massey-Ferguson)

PLATE 12.2 *Rear view of a manure spreader. The augers have shredding teeth attached to the flights. Note the left- and right-hand flights of the spreading augers. This large capacity spreader has a twin axle.*
(Krone)

two spiked shredders and an auger with left- and right-handed flights to spread the manure.

The conveyor floor is driven by a ratchet mechanism with an adjustable cam which is used to change the conveyor speed. The shredding and spreading rotors are chain driven from a gearbox on the spreader. The main drive is protected from mechanical damage by a safety slip clutch.

Most manure spreaders are constructed from steel, but some have wooden sides and flooring. The latter type can, in some cases, have the spreading unit removed leaving a useful size trailer with a self-unloading floor.

The smaller, wheel-driven models have both shredding rotors and an auger spreader. The wheels are fitted with the tread bars the opposite way to the rear wheels on a tractor. This gives the tread bars a self-cleaning action and ensures positive drive.

Manure spreaders range in capacity from 1.5–12 tonnes and have a spreading width of 2.5–12 m.

Controls

Drive engagement The power take-off lever on the tractor engages drive to the spreader. A hand lever engages the drive on wheel-driven machines.

Application rate This is altered by changing the speed of the moving floor conveyor with a lever or a screw adjuster. Tractor forward speed, changed with the gearbox, will also vary the rate of application. Reduce speed to increase the rate of spread.

Side Delivery Spreaders

The open sided cylinder shaped trailer body is suitable for handling very wet materials. The high speed flails are driven by the power take-off shaft through a chain-driven reduction unit. Before filling, the flails are wrapped around the rotor. The machine is designed to allow only the flails at each end of the rotor to spread when the spreader is full. As the load is gradually removed more of the flails are able to swing free and spread the manure.

There are a number of side delivery models on the market with capacities varying from 3–8 tonnes. The manure is discharged to one side only with a typical spread width of 3 m.

PLATE 12.3 *Side delivery manure spreader.*
(Howard Farmhand)

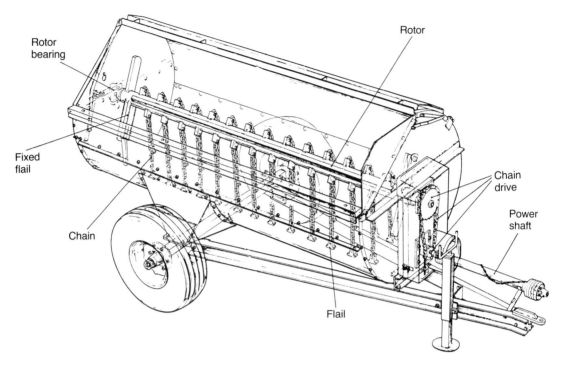

FIGURE 12.1 Sectional diagram of a side delivery spreader. (*Howard Farmhand*)

Application rates range from 12–60 tonnes per hectare (5–25 tons per acre).

Controls

Drive engagement The tractor power take-off lever engages drive to the rotor.

Application rate Tractor forward speed, changed with the tractor gearbox, varies the quantity spread per hectare. It is also possible to change the chain drive gear ratio to alter the rotor speed if tractor gear selection will not give the required dressing.

Using manure spreaders

Best results are obtained with rear spread machines by working in strips or lands. This will avoid undue idle running on the headlands. With side delivery spreaders, it is necessary to start at one side of the field and work across to the other. A tractor with a pick-up hitch makes one-man operation very easy. The driver can load his spreader with a tractor-mounted loader, then hitch up and spread the load without leaving the tractor seat.

On larger farms, one man may use the loader while a second uses two spreaders.

Spreader maintenance

All lubrication points must be greased daily during the working period. Gearbox oil levels must be checked frequently and chain drives kept correctly tensioned. At the end of the season, the machine should be thoroughly cleaned and damaged parts replaced. It is a good idea to protect the spreader from corrosion with waste oil or similar material.

SLURRY SPREADERS

Slurry can be spread on the land with special irrigation equipment using a pump, pipeline system and slurry guns which spray the liquid on to the land. More often, slurry spreaders which can load, transport and spread the material are used. Most slurry spreaders have a cylindrical-shaped tank. Capacities of 2,250–6,750 litres are in common use.

PLATE 12.4   *A medium sized slurry spreader.*
(Chillington)

Filling slurry spreaders

Vacuum filling   A power take-off-driven vacuum pump is used to suck slurry into the tank. When full, the control valve is moved so that the vacuum pump agitates the load, keeping it well mixed until it is spread.

Mechanical filling is carried out with a power-driven pump, either operated by a separate tractor with a mounted slurry pump or by a pump at the rear of the spreader. The pump conveys the slurry from the storage tank or pit into the spreader through a filler opening in the top of the tank. For very thick slurry, an auger-type conveyor may be needed.

Spreading

Vacuum-filled spreaders have a control which makes it possible to convert the vacuum pump into a pressure pump. It is then used to pressurise the tank up to about 3 bar (45 psi). This pressure is used to empty the tank when the rear outlet is opened. A deflector plate or a jet spreads the slurry in a band up to 20 m wide. Application rate is altered by changing the forward speed of the tractor and by altering the setting of the deflector plate to vary the width of spread.

Other types of spreader are emptied by gravity. The slurry passes through an adjustable opening on to a power-driven spinning disc which spreads it on to the land. Another system employs a pump to remove the slurry from the tank forcibly and spread it.

Output

Slurry tankers can run over the same area of land several times, so output is best measured in the time taken to empty and fill the machine. A typical spreader with a capacity of 4,500 litres can be both filled and emptied in 3–5 minutes.

FARMYARD MANURE LOADERS

A variety of machines can be used to load farmyard manure into trailers or spreaders. The tractor-mounted front-end loader is most common. Rear-mounted loaders and grab loaders mounted on tractors are also used. The growing use of rough terrain fork lift trucks has provided another system of manure loading. When fitted with a manure fork or bucket, these machines are capable of high work rates.

PLATE 12.5   *Slurry injection tines. Slurry from the tank passes through the tubes to furrows made in the soil by heavy tines. This system helps overcome pollution problems.*

PLATE 12.6   *Tractor-mounted loader fitted with a silage grab.*   (Massey-Ferguson)

Front-end Loaders

Most front-mounted loaders consist of two heavy duty arms (the boom) pivoted at the lower end on a frame attached to the tractor. This frame is fixed centrally between the front and rear wheels. Many attachments can be fitted to the outer end of the loader arms. These include manure fork, grain bucket, and bale loading equipment. The boom is lifted by two rams, one on each side of the tractor, which have oil pumped to them by the tractor hydraulic system. Some loader attachments have rams to carry out certain tasks such as tipping a grain bucket or gripping bales for loading on to a trailer.

There are many sizes and models of front-end loader. One suitable for a 52 kW (70 hp) tractor will have a lifting capacity of up to 1 tonne of farmyard manure.

Using front-end loaders
Loaders put a lot of strain on the tractor front wheels and axle. The front tyres should be suitable for heavy duty, usually six-ply tyres. The front tyre pressure must also be higher than that for field work. Balance weights, either on the rear wheels or to the hydraulic linkage, improve tractor stability and reduce wear on the steering linkages. Power-steered tractors are ideal for loader work.

When driving a tractor with the loader fully raised, the centre of gravity of the tractor is raised and the outfit may become unstable,

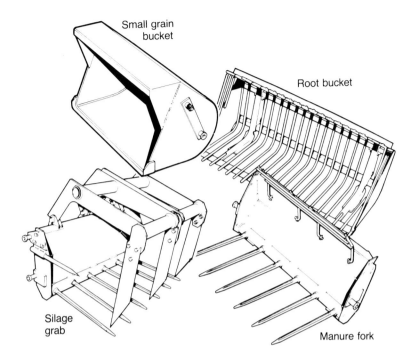

FIGURE 12.2 Forks and buckets for tractor loaders. (*Massey-Ferguson*)

especially at speed. It is best to have the loader high enough to give full vision in front without being fully raised.

To load a trailer or spreader with manure, the spreader should be set at a convenient angle to the heap to allow ease of handling. The spreader should not be set at right angles to the heap. Start loading at the front of a rear delivery spreader.

Self-propelled Loaders

Rough terrain fork lifts and skid-steer loaders are used by some farmers to handle manure. Both types of machine have a wide range of attachments available, making them suitable to handle pallets, grain, straw, building materials, etc.

Rough terrain fork lifts
The two basic types of rough terrain fork lift are the conventional rigid mast machine which has limitations in reach and lift height and the telescopic model which has a very high maximum lift.

Mast type machines may have two- or four-wheel-drive with either mechanical or hydrostatic transmission. The mast can be tilted forwards to improve reach when loading a trailer. As well as a manure loading fork, a wide range of attachments for all manner of mechanical handling is available.

These machines are rear wheel steered and have the disadvantage of carrying the load directly in front of the driver. For stability, the load should be carried as low as possible on the forks or in the bucket without obscuring the driver's view. A typical manufacturer's range of mast lift machines varies from a 2 tonne lift capacity powered by a 48 kW (64 hp) engine up to a maximum lift of 5 tonnes with a 64 kW (86 hp) engine.

Telescopic rough terrain fork lifts have a single lifting arm pivoted at the rear of the machine with the forks at the front. The telescopic loading arm has either two or three sections depending on model. It is set at one side of the machine, giving the driver good visibility. Both two- and four-wheel-drive versions are made and some

PLATE 12.7   *Telescopic loader with a manure fork. It has a 76 kW (102 hp) engine which drives all four wheels. The hydrostatic steering system can be used to steer either the front or all four wheels. Maximum lift height with the boom fully extended is 6.7 m, maximum reach is 3.55 m.*   (JCB)

PLATE 12.8   *Telescopic loader handling manure. The low boom position aids visibility and the four-wheel steering makes this large machine easy to manoeuvre in confined spaces.*   (JCB)

have the option of four-wheel steering for extra manoeuvrability. A typical telescopic fork lift has a 64 kW (86 hp) engine with a maximum load capacity of 2,500 kg which can be lifted to a maximum height of just over 9 m with a forward reach of 5.7 m. When using the maximum amount of forward reach, the safe working load on the forks must be reduced to maintain the stability of the machine.

Safety awareness is very important when using rough terrain fork lifts. The centre of gravity of the machine will rise as the load is lifted on the forks. For this reason, the load should be kept as low as possible when transporting it from one place to another. Slopes will affect the stability

of fork lifts, so it is safer to reverse down steep slopes when carrying a load. Extra care must be taken when driving across sideways slopes.

Skid-steer loaders
These machines are designed for working in confined spaces. Clearing manure from livestock buildings is a typical application. The skid-steer loader is steered by means of hand levers which lock, or partly lock, the wheels on one side while the wheels on the other side continue to drive. The result is a machine which can turn virtually within its own length. An example of a smaller machine has a 15 kW (20 hp) engine with a fork capacity of 350 kg. Much larger machines with

PLATE 12.9   *A skid-steer loader handling a forkful of manure.*   (Case IH)

45 kW (60 hp) engines or more can handle loads in excess of 1 tonne.

FERTILISER DISTRIBUTORS

Fertiliser can be spread on the land with broadcasters or full width machines. There are several types and sizes of distributor.

Broadcasters

These have a hopper for the fertiliser with a spreading mechanism below. The drive is usually from the power take-off but some machines have hydraulic motor drive. Most broadcasters are mounted but the largest capacity models are trailed. Either a spinning disc or pendulum spout is used to distribute the fertiliser.

Spinning disc models have either one or two horizontal discs which receive fertiliser from the hopper and spread it evenly on the ground. The rate of flow from the hopper to the discs can be varied to obtain the required application rate. On twin-disc machines, the drive is arranged to contra-rotate the discs, so that one disc spreads to the left and the other to the right. This gives a more even distribution pattern across the full width of spread. Mounted machines have a hopper capacity from 250 kg to 1 tonne or more. Hopper size must be related to the lift capacity of the tractor.

The effective spreading width of broadcasters varies considerably. Type of material and height of the discs above the ground are two major factors in the width of spread. A typical twin-disc machine has a spreading width of 10 m for light materials and 16 m for granular fertiliser. Another has a maximum spread pattern of 20 m.

PLATE 12.10   *Single disc mounted broadcaster.*
(Massey-Ferguson)

The largest capacity machines are trailed with a hopper capacity of up to 2.5 tonnes. Specialist lime spreaders will hold 5 tonnes or more. A conveyor belt in the hopper floor carries the fertiliser to the spinning discs at the rear. The conveyor belt is land wheel driven to maintain a constant relationship between forward speed and conveyor speed. This arrangement ensures that a consistent application rate is achieved, although ground speed may vary. These larger capacity machines are also suitable for spreading lime. Typical width of spread varies from 8–9 m for lime to up to 15 m for granular fertiliser.

The pendulum or oscillating spout spreading mechanism is an alternative to the spinning disc. The mounted or trailed hopper has an oscillating spout which distributes the fertiliser. The spout is driven by the power take-off through an eccentric unit which oscillates the spout. The rate of flow to the spout can be regulated to control application. Trailed models have a feed auger in

PLATE 12.11 *Trailed fertiliser and lime spreader with twin broadcasting discs driven by the power take-off. The cage wheel above the tyre drives the floor belt conveyor, which feeds the material to the discs. The system ensures feed is related to forward speed. This medium capacity spreader holds about 3 tonnes.*

(KRM)

PLATE 12.12 *Oscillating spout broadcaster.*

(Vicon)

the hopper floor which conveys the fertiliser to the spreading mechanism.

Pendulum spreaders range in hopper capacity from 250–1,500 kg. Trailed models have a capacity of 2 or 3 tonnes. Maximum spreading width across the range is from 6–12 m depending on type of material.

Adjustments

Application rate is controlled with an adjustable slide in the hopper bottom. This varies the amount of fertiliser passing to the disc or spout. A large opening will give a high application rate.

Forward speed will also vary the application rate. A fast forward speed, selected with the tractor gearbox, will give a low rate of spread compared with a slow forward speed.

Some machines have an adjustment to vary disc speed. This will give a wider spreading width and will reduce application rate unless forward speed is reduced or the slide is opened, to compensate for the wider spreading width.

Shut-off lever This is used to cut off the flow of fertiliser to the spreading mechanism at the end of a bout.

Attachment Once fitted to a tractor, a mounted broadcaster must be held rigid on the linkage with external check chains or stabiliser bars. It is important to set the spout or disc level with, and at the right height from, the ground. Check the instruction book for these settings. The height of the disc or spout will affect width of spread and this, in turn, can give poor matching of the work.

Using broadcasters
Equal matching of work is most important. The instruction book will give the spreading width for different application rates and materials. Powders will not be thrown as far as granular fertiliser. The only certain way of good matching is to use some form of marking aid. Stakes or empty fertiliser bags can be placed at measured intervals along the headland as a driving guide. Some farmers use markers on the tractor to drive by. Width of spread can vary from 5–20 m, according to type of broadcaster and fertiliser.

Working in strong winds will result in uneven spreading and should be avoided where possible.

Checking the application rate
The high cost of fertiliser calls for accurate application. Instruction books give a range of settings to suit all materials. Any chosen application rate can be checked, especially if the broadcaster is thought to be inaccurate. Application rate can be checked (calibrated) in the field or by a static test.

PLATE 12.13   *Roller feed broadcaster, showing the feed metering roll, the smooth upper acceleration roll and the guide channel plate.*   (Nodet)

A simple field check can be made by putting the required amount of fertiliser in the hopper for a selected area, not more than 0.5 hectare. Measure out the chosen area in a field and spread the fertiliser over this area. The hopper should be empty on completion of the measured area if the machine is accurate.

Broadcasters can also be calibrated in the farmyard. Some instruction books give the time taken to spread a stated area at a specified speed and setting. By running the broadcaster for the stated period and then collecting and weighing the fertiliser the application rate at any setting can be checked. When this information is not available, a static check on application rate can be carried out by using the method described on page 136 for crop sprayers.

Roller feed broadcaster

Fertiliser is metered from the hopper by an adjustable feed mechanism on to and through high speed acceleration rolls at each side of the machine. The fertiliser then follows guide channels and is directed in a precise spread pattern over an arc of 45 degrees to each side of the broadcaster. The two arcs meet behind the centre of the hopper to give complete coverage across a width of up to 32 m, depending on the type of material being spread.

The maximum distance of throw behind the broadcaster is approximately 4 m. The stainless steel guide channel plates can be changed to give other spreading widths if desired. The spreading mechanism can be shut off at one side, which can be an advantage when spreading the headlands.

FIGURE 12.3   Spreading pattern for a roller feed broadcaster.   (*Colchester Tillage*)

PLATE 12.14   *A large trailed pneumatic fertiliser spreader with the booms folded for transport. The cover is lifted to show the fan housing. Note the large tyres for reducing soil compaction.*   (Nodet)

Full Width Distributors

Unlike broadcasters, this type of distributor is almost as wide as the spreading width when it is in work. There are two types of full width spreader, one has mechanical feed, the other is pneumatic. The pneumatic distributor has become very popular at the expense of the various mechanical feed mechanisms.

Pneumatic distributors
The principle of pneumatic distributors is to meter, very accurately, the material to be spread into a high speed airflow created by a power take-off-driven fan. Feed rollers are commonly used to control the flow of fertiliser from the hopper. Once in the airflow the material is conveyed, along individual flexible tubes, to a number of outlets evenly spaced across the full width of the machine. Deflector plates below the outlets spread the fertiliser evenly over the ground.

A typical mounted pneumatic spreader has a working width of 12 m with a hopper capacity of about 1 tonne. There are much larger trailed machines, up to 24 m wide with a capacity of up to 6 tonnes. Application rate can be varied from, for example, 50–2,500 kg per hectare when spreading fertiliser down to as little as 5 kg per hectare when broadcasting seeds.

Application rate is controlled by adjusting the metering mechanism, usually by changing the speed of the feed rollers which meter the fertiliser into the airflow. Forward speed, varied with the tractor gearbox, will also affect the rate of spread.

Calibration can be done in the same way as for broadcasters. Some models have a calibration bag supplied which is used to gather the fertiliser. This simplifies the collection and weighing of the material.

Pneumatic distributors, like crop sprayers, have a three- or five-section boom which can be folded for transport. Folding may either be done manually or in some cases with hydraulic rams operated from the tractor seat.

Mechanical feed mechanisms
A number of machines are still in use which have mechanical feed. The distributor consists of

PLATE 12.15   *Pneumatic spreader hopper with the weather cover folded back to show the grille which prevents lumps of fertiliser clogging the mechanism.*   (Nodet)

a hopper carried on pneumatic tyred wheels with a feed mechanism in the hopper bottom. One type has a series of agitator discs running on a shaft which force the fertiliser through adjustable openings in the hopper bottom. Other full width distributors have studded roller feed similar to that used on grain drills. Fluted roller feed can also be found on some machines. Both studded and fluted roller feeds are commonly used on combined grain and fertiliser drills too.

The feed mechanism is land wheel driven so changing the forward speed does not affect application rate. This is altered by either changing the speed of the feed mechanism, usually with some

form of gearbox, or by changing the size of the fertiliser outlets in the hopper with a hand lever.

Using full width distributors
Pneumatic spreaders have an even spread pattern across their width, so careful matching of bouts is important. This can be achieved either by measuring and placing marking stakes at the correct intervals across the field or by using a foam blob marker. This is an attachment which deposits blobs of foam on the ground from discharge spouts at each end of the spreader boom. On the next run, the driver must keep the end of the boom matched with the foam

PLATE 12.16 *Pneumatic fertiliser spreader boom: the tubes from the blower and metering device can be seen with the deflector plates at each outlet point.*
(Nodet)

ried out by using the same method as that for grain drills.

Maintenance
Daily lubrication is important and careful cleaning after use is essential to ensure a long working life. All traces of fertiliser must be removed to prevent corrosion. When using a hose pipe or pressure washer, make sure that water is not left in the machine. After washing a pneumatic spreader, run the machine to let the fan blow any trapped water away.

Check that all drives, especially vee-belts, are correctly tensioned. When storing the spreader after use it is advisable to coat any bright parts, which come into contact with fertiliser, with a rust preventative solution or oil. However, do not let oil touch rubber parts or get into the air tubes.

Plastic is used extensively in the manufacture of fertiliser spreaders, which has considerably reduced the corrosion problem.

SUGGESTED STUDENT ACTIVITIES

1. With the aid of the instruction book, locate all controls and adjustments on one type of manure spreader.
2. Find out how to operate the controls of a tractor mounted manure loader.
3. Set the application rate on a fertiliser broadcaster. Refer to the rate charts in the instruction book.
4. Trace the flow of fertiliser through a pneumatic spreader.

SAFETY CHECK

When parking a fork lift truck or a tractor with a loader, always lower the fork, bucket or other attachment to the ground.

marks. Tramlines in cereal crops make bout matching simple, provided that the spreading width matches a multiple of the sowing width of the drill.

Mechanically fed distributors are matched by using the mark made by the wheels. No other form of marker is necessary. Calibration is car-

Chapter 13

GROUND CROP SPRAYERS

Crop sprayers are used to apply chemicals, diluted with water, to the soil or to growing crops for the control of insects, weeds and fungal diseases. They are also used to apply liquid fertilisers and plant growth regulators. Sprayers may be self-propelled, trailed or mounted. Spraybar widths vary from 7–24 m, with either hydraulic- or atomiser-type nozzles.

TYPES

Self-propelled sprayers
With tank capacities ranging from 500–3,500 litres, these are popular on large arable farms and with agricultural contractors. Small self-propelled sprayers often take the form of a sprayer unit mounted on a farm pick-up truck. Wide wheels and tyres are fitted to reduce soil compaction. The sprayer unit can be removed at the end of the spraying programme, leaving the truck free for other work.

A typical farmer-owned self-propelled sprayer will have a tank capacity in the range of 1,500–2,000 litres. It has the advantage of low ground pressure which reduces soil compaction; also, a tractor is not tied up for the season and the sprayer is ready for use at any time when conditions are suitable. The driver is seated in a forward control cab with excellent visibility of the crop. Some self-propelled sprayers have high clearance axles and wheels making them ideal for working in tall crops.

Trailed sprayers
These have tanks holding 1,500–3,500 litres and are found on many large arable farms. Spraybar widths vary from 9–24 m. Their main advantages are the large capacity tank and simple attachment to the tractor.

PLATE 13.1   *Self-propelled sprayer working on tramlines.*   (Sands)

PLATE 13.2   *A lightweight crop sprayer, suitable for small farms.*   (Lely)

Mounted sprayers
These are made in many sizes, ranging from small, simple machines with a tank capacity of about 600 litres to the more sophisticated models with a tank which holds 1,300 litres or more. Small sprayers with a 200-litre tank and a 6-m spraybar are made for tractors in the 12–15 kW (16–20 hp) range. Most mounted sprayers have an adjustable spraybar which enables the driver to set it in a high position for spraying in standing crops, especially cereals.

Traditionally, sprayers have been grouped according to the amount of liquid applied per hectare (acre). Modern chemicals are generally applied at fairly low rates, so most sprayers can handle the full range of chemicals. Low volume application is up to 250 litres per hectare (20 gallons per acre), medium volume from 250–500 litres per hectare and high volume up to around 1,200 litres per hectare (100 gallons per acre).

Very low application rates, as little as 25 litres per hectare (2 gallons per acre), have become popular in recent years. This reduces the amount of water to be carried, gives more hectares per tankful and cuts the cost of chemicals. The very low rates can be achieved with rotary atomisers instead of the normal spray nozzles. The atom-

isers produce uniform, medium size droplets which make very effective use of the chemical.

The Parts of a Sprayer

Tank
This is usually made of non-corrosive plastic material such as polyethylene or reinforced fibre-glass. Some sprayers have an extra tank on the front of the tractor, which reduces the time spent travelling from farm to field. A transfer pump is used to move the liquid to the main tank. It also keeps the contents of the front tank well mixed while spraying the chemical from the rear tank.

Pump
Power take-off driven, it may be of the roller vane, piston or diaphragm type. Roller vane pumps are little used now because of corrosion and lubrication problems. Multi-diaphragm pumps are in common use. Pump size will vary with type of sprayer. Typical pump outputs are:

- 1,000 litre mounted sprayer with a diaphragm pump—150 litres per minute.
- 2,000 litre trailed model with a six-cylinder diaphragm pump—230 litres per minute.
- 2,500 litre self-propelled model with a six-cylinder diaphragm pump—300 litres per minute.

Sprayer pumps should run at the standard 540 rpm power take-off speed. Operating pressures vary from 1 to 6 bar or more, but high pressures produce very small droplets which can cause spray drift. This must be avoided, especially when working in windy conditions.

Relief valve
An adjustable relief valve is included in the pump circuit to control working pressure. It is one of the factors which determines the application rate. The relief valve also returns excess chemical, not required by the nozzles, back to the tank, and agitates the tank contents, keeping them well mixed.

Filters
These are a vital part of the sprayer circuit. The nozzles have very small holes which will easily block if particles of dirt are allowed to reach them. The location, and number, of filters varies with different sprayers.

PLATE 13.3   *A self-propelled sprayer with four-wheel-drive with the advantages of high ground clearance and low ground pressure. The machine shown has a 45 kW (60 hp) engine and a top speed of almost 31 km/h. The sprayer unit can be removed and a fertiliser spreader fitted in its place.*   (Lely)

There are usually four stages of filtration. The first is a filter basket in the filler, the second is in the suction line before the liquid enters the pump, the third is in the delivery line from the pump and final filtration is frequently included in each nozzle. Sprayers with self-fill hoses will have a filter unit at the end of the suction pipe. All filters must be kept clean.

Mixing devices

As an aid to safe chemical handling, some sprayers have a chemical suction probe operated by the sprayer pump. After placing the probe in the container the required amount of chemical for the mix is sucked into a small measuring tank on the side of the machine. The chemical is then transferred to the water in the main sprayer tank and thoroughly mixed by pumping the contents round the system.

Powdered chemicals can be pre-mixed with a small quantity of water in a mixing bowl which is fitted to some sprayers. After the powder and water are well mixed, the concentrated chemical is transferred to the water in the main tank and thoroughly mixed.

Mixer tanks

Travel time to and from the farm when spraying

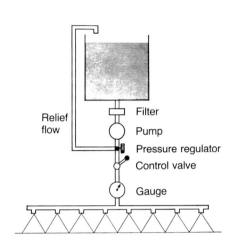

FIGURE 13.1   Simple crop sprayer spray circuit.

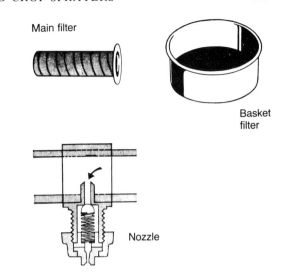

FIGURE 13.2   Three stages of filtration.

can be reduced by using a mixer tank. This is a large water tank on its own wheels complete with chemical mixing equipment. The mixer tank is filled with water, then the correct quantity of chemical is added and mixed with a power take-off or engine-driven pump. The filled mixer tank is then towed to the field where the mixed chemical is pumped into the sprayer tank with a flexible hose pipe.

Controls
The control unit is operated either manually with a lever, especially on older sprayers, or electrically from the tractor cab. Like many other farm machines, control systems vary with make and model. The basic function of the control unit is to switch the sprayer on and off. When switched off, the chemical is sucked back from the nozzles to prevent drips and the pump circulates the chemical around the system to keep the tank contents mixed.

Spray pressure is regulated by adjusting the pressure relief valve. Some sprayers have an automatic control system which regulates the spraying pressure according to ground speed, thus maintaining the correct output. The control needs readjustment if a different gear is selected.

Many sprayers have a self-filling hose which is operated from the main control unit. It is used to pump water from a mobile tank, which saves making journeys to and from the farm.

Spraybar
The spraybar and nozzles are made of non-corrosive material, usually plastic, which is supported by a light, but strong, metal frame. It has three or five sections which are folded for transport, either by hand or hydraulically. Spraybar widths of 7–24 m are in common use; vertical adjustment is provided to suit different crop heights.

PLATE 13.4   *This mixer tank is filled and emptied by a power take-off driven pump in about five minutes. The small tank on the side is used to mix powders and liquid chemicals and transfer them to the main tank for mixing with the tankload of water.*   (Lely)

PLATE 13.5 *The small tank with a conical bottom is the chemical mixing bowl on this mounted sprayer.* (Vicon)

A major problem, especially when working on rough ground, is bouncing and swaying of the spraybar. This movement alters the position of the nozzles in relation to the spray target, causing overdosing or missed strips. On some machines the spraybar is attached to the frame by means of a stabilising linkage. This absorbs the movement of the tractor on uneven surfaces and in so doing cuts out sway and bounce to keep the spraybar parallel with the ground.

A *break-back device* is built into the spraybar hinge points. This allows the bar to swing backwards, preventing damage if it hits an obstruction.

Isolating taps are provided which enable the driver to turn off a section of the spraybar. This is useful, for example, when less than a full spray width is required or a strip around the edge of the field requires treatment.

Nozzle

The nozzle body is made of plastic or metal, but the nozzle itself may be plastic, brass or ceramic. More than one set of nozzles will be required to obtain the full range of application rates. The two common types of nozzle give a flat fan and a cone-shaped spray pattern.

Many sprayers have an anti-drip device in each nozzle which prevents nozzle drip when the machine is turned off. One type consists of a small spring-loaded valve which seals the nozzle outlet when the sprayer is switched off. Another anti-drip device consists of a small diaphragm which closes the nozzle outlet when spraying ceases. Some sprayers have multi-head nozzles which can be rotated when a change of jet size is required.

Working Adjustments

Application rate

Three factors affect the application rate of crop sprayers:

- *Tractor forward speed* A low forward speed gives a high application rate. By changing gear to double the forward speed, the application rate will be halved.
- *Spraying pressure* Higher application rates are obtained by adjusting the relief valve

PLATE 13.6 *A trailed sprayer with lightweight aluminium booms attached by a special suspension system designed to overcome the problems of spraybar bounce and sway usually experienced on uneven surfaces.* (Lely)

Fan jet

Hollow cone jet

Anti-drip device in nozzle

FIGURE 13.3 Types of nozzle.

PLATE 13.7 *Five-way nozzle giving instant jet-size selection by rotating the unit.* (Technoma Sprayers)

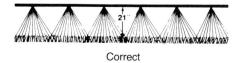

Correct

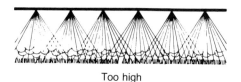

Too high

Too low

FIGURE 13.4   The effects of incorrect spraybar height.

control to increase working pressure. High pressure reduces droplet size and increases the risk of the chemical being blown away by the wind (drift). This should be avoided as drift can damage nearby crops and gardens. Spray drift is also a serious threat to the environment as it will harm trees, hedgerows, wild flowers, etc.

• *Nozzle size* Nozzles with a large hole will give a high application rate. Smaller ones must be fitted when a reduction in application rate is required.

Once set, the application rate can be kept constant with an automatic control system. This electronic device is found on the more expensive sprayers. Some sprayers have a system of pressure compensation which avoids variations in working pressure when a section of the spraybar is turned off.

Spraybar height
The spraybar must be level when in work. The spray pattern from each nozzle should meet just above the target. When spraying weeds which are taller than the crop, the weeds must be considered the target. On most sprayers, the nozzles are spaced at 500 mm intervals along the spraybar. The height of the nozzles above the target is determined by the angle of the spray pattern. For example, nozzles at 500 mm spacing with an 80 degree fan or cone pattern should be 530 mm above the target; a 110 degree fan pattern should be 360 mm above the target. When weather conditions allow, the nozzles can be set a little higher to avoid missed strips when uneven field surfaces cause spraybar bounce. The height of the spraybar is adjusted by moving

PLATE 13.8   *Crop sprayer with spinning disc atomisers.*   (Technoma Sprayers)

it up or down on the frame. This adjustment is important because incorrect nozzle height will cause overdosing or missed strips.

Controlled Droplet Application (CDA) Sprayers

Low volume application rates can be achieved by using rotary atomisers instead of normal nozzles; 25 litres per hectare (10 gallons per acre) is a typical application rate. Rotary atomisers are suitable for most farm chemicals. CDA sprayers produce droplets of uniform and medium size, thus avoiding the very small droplets which drift or the very large ones which tend to run off the leaf. The atomisers are spaced at intervals of 1–1.5 m on the spraybar, depending on model. There are two types of rotary atomiser:

Horizontal atomisers are driven by small hydraulic motors powered by the tractor auxiliary hydraulic supply. Chemical is pumped from the tank to the atomisers. Uniform droplet size is achieved by feeding chemical to the bottom of a fast spinning cup in the atomiser and allowing it to be spun out under gravitational force. The cup has a large number of grooves on the inside, which guide the chemical evenly to the outlet. An automatic pressure control system maintains the correct supply of chemical to the atomisers, although there may be variations in power take-off speed.

Vertical spinning discs give a spray pattern similar to that of a flat fan jet. Chemical is fed to the centre of a vertical spinning disc, about 150 mm in diameter, which is then thrown outwards by gravitational force giving a narrow band of droplet sizes. Each disc is driven by a 12 volt electric motor supplied with current from the tractor battery. It is possible to change the size of the droplets but still maintain their uniformity by changing the speed of the spinning discs. This can be varied from 1,100–4,200 rpm, the higher speeds giving the smaller droplet size.

Advantages of CDA include more hectares per tankful, and uniform sized droplets giving better chemical retention on the plants with less drift and run off.

One model of sprayer has both hydraulic nozzles and rotary atomisers on the same spraybar. It is a simple task for the driver to convert from one system to the other, depending upon weather conditions and the chemical to be used.

PLATE 13.9 *Rotary atomiser.* (Lely)

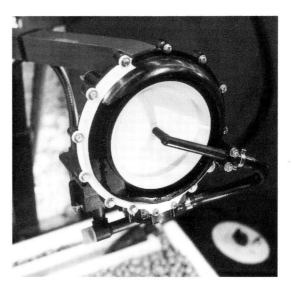

PLATE 13.10 *Close-up of a spinning disc rotary atomiser. At a forward speed of 8 km/h it will apply from 2.5–5 litres per hectare depending on setting. The disc is driven by an electric motor, and has six speed settings which vary the size of the droplets.*
(Technoma Sprayers)

Twin-fluid nozzle sprayer
One of the latest developments in crop sprayer design is a special type of nozzle with both liquid and air feed. The liquid is conveyed to the nozzles by a conventional sprayer pump. A high volume, low pressure air supply is produced by a compressor on the sprayer. The air and liquid are mixed within the nozzle and forced through a small hole on to a deflector plate to produce a flat fan-shaped spray pattern.

Using Sprayers

Before the season starts

- Repair any damaged or worn parts.
- Turn the pump by hand to check it has not seized.
- Hitch the sprayer to the tractor and connect the pump to the power take-off shaft.
- Clean the inside of the sprayer tank, remove the nozzles and then flush the machine with a tankful of clean water.
- Wash and replace the filters, fit a clean set of nozzles and then spray out another tank of clean water, checking the spray pattern.
- Make sure all the nozzles have the same output. To do this, run the machine, and collect water from each nozzle for half a minute. By using a measuring jug to collect the water, the exact output of each nozzle can be obtained. Replace any unsatisfactory nozzles. Nozzles will wear with use and will need replacing after a time.
- Finally, check the sprayer for leaks.

Spraying
Set the sprayer to give the required application rate. Fit the correct nozzles, then adjust to the specified pressure while spraying clean water.

Select the correct forward gear, to give the required speed, with the engine revolutions set to give 540 rpm at the power take-off.

Read the instructions on the chemical container. Make sure you use the correct protective clothing for the chemical you are spraying.

Partly fill the tank with water, add the chemical in the way indicated on the container, then add the required quantity of water to complete the tank mix. Do not remove the filter basket in the filler opening to speed up the procedure. After filling, agitate the contents thoroughly by pumping the tank load round the system. It will then be ready for spraying. Many sprayers have a mixing bowl, which makes filling the tank with chemical and water a much safer procedure.

Once in the field, spray round the headlands twice, then start on the main part of the field, working from one side towards the other. Direction of spraying will depend on wind direction, neighbouring crops and field shape.

To help with matching the work, a long length of stout twine can be tied on to each end of the spraybar. When turning on the headland, the end of the twine will stay in position until the spraybar passes over it on the return run. A foam-marking device which drops blobs of foam from the end of the spraybar or marking stakes can be used as a guide. In some crops, the rows themselves provide the driving guide.

A tramline attachment on a grain drill provides a permanent driving guide by leaving two undrilled rows at selected intervals. The tractor wheels run on the undrilled land for all fertiliser and spray treatments during the life of the crop. For full benefit, the sprayer and the fertiliser distributor must have working widths which are multiples of the drilling width. The tramline system ensures minimum crop damage together with accurate crop protection.

Checking sprayer application rate
Accurate application of spray chemicals is most important. A sprayer can be calibrated by static or field methods.

Static test This can be done by running the sprayer for a calculated time, equivalent to spraying one hectare. Use this formula to find the time to spray one hectare:

$$\text{Time (mins)} = \frac{600}{\text{Spraying width (m)} \times \text{speed (km/h)}}$$

For example, time to spray one hectare with a spraying width of 10 m at 6 km/h will be:

$$\frac{600}{10 \times 6} = 10 \text{ minutes}$$

Select the required nozzles and pressure. Fill the tank with water and run the sprayer for the calculated time. Then refill the tank, measuring the quantity of water needed. This amount will be the application rate at the chosen setting.

Field test Calibration in the field is done by driving the outfit for a calculated distance, using the forward speed, pressure and nozzle size to give the chosen output. Use this formula to find the distance to travel to spray 0.5 hectare:

$$\text{Distance (m)} = \frac{5,000}{\text{Spraying width (m)}}$$

To find the spraying width, multiply the

number of nozzles by the nozzle spacing. For example, if a sprayer has 25 nozzles spaced at 0.5 m, this gives a spraying width of 25 × 0.5 which equals 12.5 m:

$$\text{Distance to travel (m) for 0.5 hectare} = \frac{5,000}{12.5} = 400 \text{ m}$$

With a full tank, drive the calculated distance, then refill the tank, measuring the amount of water needed. This will be the application for 0.5 hectare, so the quantity must be doubled to find the application rate per hectare.

Nozzles will wear with constant use. Their output should be checked at the start of the season. With plenty of water in the tank, run the sprayer and collect water, for a fixed time, from each nozzle in turn. Use a calibrated measuring jug so that it will be easy to determine the exact amount of water delivered by each nozzle. Half a minute running time for each nozzle should be sufficient. Replace any nozzles which have a high or low output. Make sure that the nozzle and its filter are clean before checking output.

It is important that the forward speed is accurate when spraying. This can be checked by measuring the time taken for the tractor to travel 100 m at any selected engine speed and gear. Use this method:

1. Accurately measure a distance of 100 m and mark with a stake at each end.
2. Select a gear which gives the required forward speed and set the throttle to give a power take-off speed of 540 rpm.
3. Carefully measure the time taken to travel 100 m.
4. Use this formula to calculate the speed:

$$\text{Speed in km per hour} = \frac{360}{\text{Time in seconds}}$$

$$\text{e.g. Speed} = \frac{360}{60 \text{ seconds}} = 6 \text{ km per hour}$$

5. Keep a record of the gear used, engine speed and forward speed for future reference.

To calculate forward speed in mph, the distance to travel is 100 yards and the speed is calculated by dividing 205 by the time taken in seconds, e.g. the time to travel 100 yards is 68 seconds; the speed will be 205 ÷ 68 = 3 mph.

Sprayer maintenance

At the end of the day Empty the tank of chemical, then partly fill with water. Clean the machine, both inside and outside, remembering to wear protective clothing. After cleaning, spray out the washing water on waste land. Never dispose of tank washings where they could harm humans, animals, fish or other crops. Finally, remove, clean and replace filters and nozzles.

At the end of the season Remove all traces of chemical from the sprayer. Clean the nozzles and store them separately. Drain the pump to prevent frost damage. Release all pressure on the relief valve and, after repairing any damage, store the sprayer under cover.

Cleaning the sprayer When changing from one chemical to another, or when preparing for storage, all traces of chemical must be removed. Small residues of chemical can damage other crops at a later date.

For thorough cleaning:

- Remove the drain plug and empty the tank.
- Scrub the machine with a stiff brush, especially the underside of the tank top. Hose down the sprayer.
- Replace the drain plug. Fill the tank with clean water and add a cleaning agent. Use liquid detergent if the chemical contained oil, otherwise use washing soda. Read the spray chemical instructions. Circulate the cleaning water through the sprayer with the pump, then spray out. Repeat the process a second time.
- Rinse the machine by spraying out two tanks of water.
- Remove, wash and replace all filters and nozzles.

SAFETY WHEN SPRAYING

Regulations were introduced in 1986 concerning the safe use of pesticides. Under the Regulations, the term pesticide refers to all types of agricultural chemical such as herbicides, fungicides, plant growth regulators and sterilants. Application of pesticides by hand-held equipment, tractor sprayers and aircraft are among the list of chemical applicators covered by the Control of Pesticides Regulations 1986.

PLATE 13.11   *A tractor equipped with flotation tyres and a large capacity sprayer with an extra tank at the front. The large tyres are used to overcome the problems of soil compaction caused by heavy farm equipment.*

(Massey-Ferguson)

All people who handle or apply pesticides must have proof that they have received adequate instruction in their use. Certificates of Competence are required by any person applying pesticides on property not in their occupancy or the occupancy of their employer. Anyone applying pesticides who was born after 31 December 1964 also requires a Certificate of Competence.

The information given above is very basic and is only an indication of the content of the Control of Pesticides Regulations. You should study these Regulations to find out exactly how they affect you when handling or using pesticides.

Some important points to remember when handling or applying agricultural chemicals include:

- Always read the instructions on the package first. Use protective clothing as directed on the label.
- Check the label for any special washing requirements to be followed after using the product.
- Dispose of empty containers safely. Wash out cans to remove any undiluted chemical.

- Keep all chemicals in a locked store. Return any unused material to the store.
- Do not smoke, eat or drink while spraying. Always wash before doing any of these.
- Never blow at a blocked nozzle with the mouth. Keep spare clean nozzles with you. (Do not try to clear a blocked nozzle with a pin either—this will spoil it.)
- Never transfer chemicals into other containers.
- Clean all protective clothing after use.
- Remember that spray chemicals can enter tractor cabs. A respirator is needed, even in modern safety cabs, for some of the more dangerous chemicals.

BAND SPRAYERS

These are very similar to ordinary crop sprayers but apply chemicals in narrow bands, saving expensive chemicals where overall coverage is unnecessary. Band sprayers are mostly used with precision seeders. Carefully calibrated and matched nozzles are used to apply pre-emergence weed killers when drilling such crops as sugar beet. The young plants can then emerge in a weed-free band, making inter-row hoeing an easier task.

The sprayer tank is usually carried on the tractor toolbar but sometimes on the front of the tractor. The control unit directs the chemical to nozzles attached to the back of the seeder units.

Band sprayers are also used to apply weed killers and pesticides to growing crops.

GRANULAR APPLICATORS

Crop protection chemicals can also be applied in a solid, granular form. Granular applicators will apply these granules to growing crops or in bands when planting potatoes and other seeds.

The granules are carried in small hoppers attached to a toolbar. A mechanical or pneumatic feed mechanism delivers granules to nozzles spaced across a toolbar or fixed to coulter units. The flow of granules from the hopper can be varied to give a range of application rates.

SUGGESTED STUDENT ACTIVITIES

1. Locate all filters on a crop sprayer.
2. Identify the various crop sprayer controls.
3. Look for a sprayer with rotary atomisers, perhaps when visiting an agricultural show. Study the construction of an atomiser unit and how it is driven.
4. Look for the types of chemical mixer described in this chapter.
5. Read the instructions on spray chemical cans and note the protective clothing required.

SAFETY CHECK

Make sure that you comply with the Pesticide Application Regulations. You must have the proper training and probably need a Certificate of Competence when using pesticides. Find out which materials come within the scope of the Pesticides Regulations.

Chapter 14

HAYMAKING MACHINERY

MOWERS

Mowers are used to cut grass and other forage crops. The cut material is left in a swath for further treatment in the process of making hay or silage. For many years, mowers were of the cutter bar type. However, these machines have now been replaced by the rotary mower and to a lesser extent by the flail mower.

Rotary Mowers

Rotary mowers have a high work rate and relative freedom from blockages, even in heavy or tangled crops. Working speeds of up to 15 km/h (9 mph) can be achieved in good conditions.

Either mounted on the three point linkage or trailed, rotary mowers are driven by the power take-off. Some are front mounted, leaving the

PLATE 14.1 *Trailed mower conditioner cutting and treating the swath in one pass.* (Vicon)

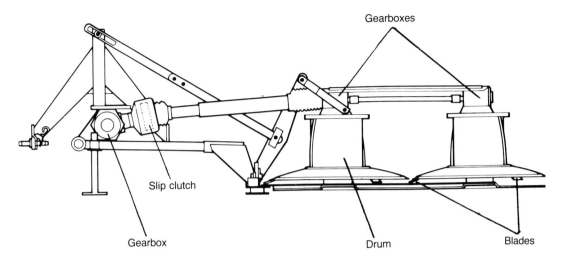

FIGURE 14.1   Drum mower.   (*Massey-Ferguson*)

rear linkage free for swath treatment machinery. Most rotary mowers have a cutting width within the range 1.4–3.2 m. Some machines have two or three large diameter cutting rotors or drums with from three to six cutting blades similar to those used on a rotary lawnmower. Others have much smaller diameter discs with two or three cutting blades on each.

Drum mowers

The large diameter rotors of a drum mower are driven from the top, either by a system of gears or a toothed belt. Positive drive is necessary to prevent the replaceable blades on adjacent drums coming into contact with each other. The rotors turn at speeds in the region of 1,000 rpm.

The drive system is arranged so that, for

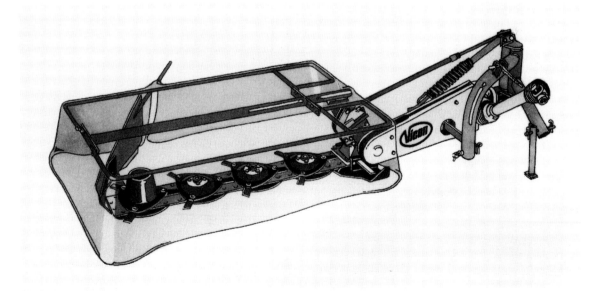

PLATE 14.2   *Cutaway view of a disc mower. The outer disc has a drum which, assisted by the swath board, moves the crop to leave a clear space for the tractor wheels on the next round.*   (Vicon)

a two-drum model, the rotors contra-rotate to form a single swath behind the mower. A three rotor machine leaves one full swath and one half swath which is completed on the next cutting bout. A tractor of about 40 kW (55 hp) is needed to operate a typical twin rotor mounted mower with a cutting width of 1.85 m.

Disc mowers

This type of mower is also gear driven to prevent the blades on adjacent cutting discs coming into contact with each other. The discs, with two or three replaceable blades, depending on model, are carried on a combined cutter bar and gear housing which runs along the ground. Skids at each end of the bar prevent excessive wear on the underside of the housing containing the gears which drive the discs at about 3,000 rpm. Each pair of discs contra-rotates so that, for example, a four-disc model leaves two swaths. The discs at each end of the bar have a steel cone which acts as a deflector to keep the cut material well away from the standing crop. A 40 kW (55 hp) tractor will be needed to drive a four-disc mower with a 1.65 m cutting width.

The high cutting speed of rotary mowers will cause stones to be thrown from the rotor with considerable force. A protective shield is fixed above, behind and at each side of the rotors to protect the operator and bystanders from injury. Never walk behind a rotary mower when it is cutting, or allow others to do so.

The cutter bar lift linkage has a balance spring which, when correctly adjusted, helps the discs or rotors to follow uneven field surfaces and

PLATE 14.3 *Detail of a disc mower cutter bar showing the arrangement of the discs so that the blades do not collide.* (Krone)

maintain the correct cutting height. The bar will fall heavily when lowered into work if there is too little tension on the balance spring.

Controls and adjustments

Drive engagement The power take-off lever is used to engage drive to the cutting mechanism.

Cutter bar position The three settings are controlled by the hydraulic lever:

1. *Transport* For moving from field to field, the mower is lifted hydraulically and then raised to an upright position either with a crank handle or an auxiliary hydraulic ram. Some models are transported by swinging the cutting unit round behind the tractor to give a narrower overall width.
2. *Turning* The cutter bar is raised hydraulically for turning on the headland. The outer end of the bar automatically lifts higher than the inner end to stop it catching on the previously cut grass.
3. *Cutting* The bar is lowered with the hydraulic system; the actual cutting height is set with skids, one at each end of the bar. Topping skids can be fitted to some models. They raise the cutting height by about 100 mm for topping pastures.

Trailed rotary mowers have an auxiliary hydraulic ram for lifting and lowering the cutter bar.

The balance spring must be adjusted to help the bar follow uneven field surfaces.

Pitch The top link is used to adjust cutter bar pitch. The bar will normally run level but, by shortening the top link, the cutting blades are brought even closer to the ground to give a very close cut. This should not be done on stony fields.

Maintenance

The cutting blades must be kept sharp and in good condition. They are easy to change when worn.

Belt tension must be checked at regular intervals. Most belt-driven mowers have three or four vee-belts, side by side on multiple vee-belt pulleys. The belts must always be replaced in sets. When a broken belt is replaced with a new one and the older unbroken belts are left

PLATE 14.4 *Disc mower attachment to a tractor. The balance spring which helps the bar float over uneven ground and the ram used to raise the bar to a vertical position are shown.* (Krone)

in place, the old belts will not transmit very much power. This is because, when the new belt is tight, the old belts will still be only partly tensioned.

Gearbox oil levels should be checked at regular intervals. Lubricate the mower every day during the season. Watch for oil leaks, especially from the gear case on bottom-driven mowers as the casing runs close to the ground and may suffer damage in rough conditions.

Safety devices

Break-away device A spring-loaded mechanism allows the whole cutter bar to swing backwards if it hits an obstruction. Reverse the tractor to return the cutting mechanism to the working position.

Slip clutch Some rotary mowers have a slip clutch in the drive from the power take-off shaft to the mower gearbox. This protects the drive from overload.

Shear bolt Some models have a shear bolt breakaway device. If the bar hits an obstruction the shear bolt breaks, allowing it to swing backwards, avoiding damage.

Flail Mowers

The flail mower has a high speed rotor, fitted with swinging flails which cut the grass and leave it in a fluffy swath. The power take-off-driven rotor is enclosed in a metal shield which helps to form the swath. The flails bruise the grass as it is cut. This speeds up the haymaking process but can cause some loss of material.

The drive from the power take-off is trans-

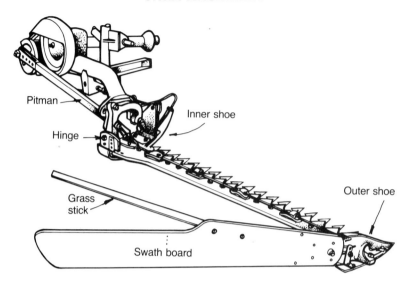

FIGURE 14.2   Cutter bar mower.

mitted through a gearbox and either a belt or chain drive to the rotor.

Some flail mowers are off-set. This means that the cutting mechanism is to one side of the tractor so that the tractor wheels do not run on uncut crop. Others are in-line, with the cutting rotor running directly behind the tractor.

Cutting height is adjusted with skids at each side of the rotor housing. Off-set flail mowers are transported with the cutting unit folded back behind the tractor.

Cutter Bar Mowers

The reciprocating knife cutting mechanism is similar to that of a combine harvester. The cutter bar inner shoe is hinged to the mower frame, which is attached to the tractor hydraulic linkage. The knife is driven by the power take-off through a crank and pitman (connecting rod). The cutting action is described in Chapter 17. The outer shoe on the cutter bar (see Figure 14.2) has a swath board which deflects the grass to leave a space for the tractor wheels on the next round. Cutter bars are 1.5 or 1.8 m long. The drive crank runs at approximately 500 rpm, giving about 1,000 cutting strokes per minute. The knife stroke is 75 mm. Safety devices include a cutter bar break-back mechanism and a slip clutch in the main drive.

Controls and adjustments

Drive engagement   The power take-off lever is used.

Cutter bar position   The bar can be set in the same three positions as those described for a rotary mower. The bar is raised to the upright position by hand and held in position with a stay when transported from field to field.

Cutter bar pitch   An adjuster on the inner shoe is used to change the angle of the cutter bar fingers in relation to the ground.

Height of cut   Adjusted with skids on the inner and outer shoes.

Cutter bar balance   An adjustable balance spring must be tensioned to allow the bar to follow uneven field surfaces and maintain the correct cutting height.

Maintenance

The knife must be kept sharp at all times. The smooth edged sections should be sharpened with a file or knife grinder. Damaged sections should be replaced and the knife back must be straight. Never oil the cutting mechanism, except when preparing it for storage. The cutter bar must have lead: this means that the outer end of the bar is set forward of the inner end by about 6 mm for every 300 mm of cutting width.

Using mowers

The headland must be cut first. It must be wide enough to allow easy turning when cutting the main part of the field, either round and round or in sections (lands). The actual cutting plan will depend on the shape and size of the field. Rotary mowers can be used at high speed in all but the most tangled crops. Cutter bar mowers do not have the advantage of high speed work; this, together with a tendency for the cutter bar to block in heavy crops, is a reason for their loss of popularity.

MOWER CONDITIONERS

Freshly cut grass will have a moisture content of about 75 per cent. For safe storage in the hay barn this must be reduced to approximately 20 per cent. It is important to reduce the moisture level as quickly as possible with the help of the sun and wind. Hay treatment machines help the drying process by turning and loosening the hay swaths. This work must be done with minimum loss of the valuable leafy material.

Mower conditioners combine cutting with the first treatment of the new swath. The cut grass passes to the conditioner which bruises the stem to release the sap and speed the drying process. There are several types of conditioning mechanism. Some conditioners have free swinging nylon fingers on a rotor shaft which treat the whole crop as it leaves the cutting discs.

PLATE 14.5 *Off-set trailed mower conditioner. The mower wheels run with one at each side of the conditioned swath.* (Massey-Ferguson)

Another system uses a horizontal rotor which has vee-shaped spring tines: these scuff the grass stems to help release the sap as the crop leaves the cutting discs or drums.

A third type of conditioner consists of two intermeshing nylon rollers held in close contact by spring pressure. The surfaces of the rollers have grooves which mesh with each other. As the grass passes between the rollers the stems are crimped to release the sap and hasten drying.

The power take-off driven conditioner rotor is contained within a metal or plastic hood with rear doors. The doors can be adjusted to restrict the outflow of the crop when a greater degree of conditioning is required. The width of the conditioner will be matched to the cutting width of the mower. Both mounted and trailed models are in use.

Conditioners are also available as separate machines with either a drawbar or hydraulic linkage and power take-off drive. These machines can be used in conjunction with a mower. The mower is usually front or side mounted with the conditioner at the back.

HAYMAKING MACHINERY

Haymaking machinery is used for a variety of treatments:

- *Tedding* The tedder is used to lift and loosen the swath, enabling air to circulate freely through the grass.
- *Swath turning* This process involves moving individual swaths sideways on to fresh ground and at the same time turning them over.
- *Side raking* Two or more swaths are combined to form a much larger one in preparation for baling.
- *Spreading* The swath is spread over a wider area to reduce the depth of material and hasten drying.

Rotary Tedders

There are a number of designs, either mounted or trailed. Drive is from the power take-off.

The rotary tedder has one or more power take-off driven rotors, with spring tines, which rotate on a vertical axis. One model has a large diameter rotor with a working width of

PLATE 14.6 *A disc mower and flail conditioner combination. Three bladed discs cut the crop which then passes over the flail type conditioning rotor. It is guided back to the ground by the adjustable rear doors. The cutting discs run at 3,000 rpm and the nylon flails at about 1,000 rpm. The machine is lifted from work by hydraulic rams. Large coil suspension springs and guide skids at each side ensure that the discs follow the ground contours.*   (Vicon)

3 m; another has two rotors, again with a 3 m working width. An example of a large tedder has six rotors which cover a width of 6.7 m. When in work the tines lift and loosen the swath as it is moved sideways on to fresh ground. Many rotary tedders can be used for spreading, swath turning and side raking as well. This is achieved by reversing the drive of one or more rotors so the swath is moved in a different way.

Overshot Tedders

These machines have a horizontal rotor with a series of spring tines which lift the crop over the top of the rotor, then return it to the ground in a loose swath. The rotor runs under a hood which

has adjustable deflector doors at the rear. The shape and size of the swath is controlled by the position of these doors.

Finger Wheel Turners

Floating finger wheels carried on a hydraulically mounted frame are driven by ground contact. The wheels have spring tines which move the swath sideways on to fresh ground when used as a swath turner. The floating action allows the wheels to follow unlevel field surfaces. Finger wheel turners may have between four and ten wheels, attached to two or more adjustable frames, depending on working width. The arrangement of the finger wheels can be changed

PLATE 14.7   *A rotary tedder. The tines move outwards when the machine is in use. The direction of rotation is altered on one or both rotors to achieve different treatments to the swath.*   (Massey-Ferguson)

to make the machine suitable for spreading or side delivery raking. This type of turner can be used at speeds of up to 22 km/h (14 mph). In addition to rear-mounted finger wheel turners, both front-mounted and trailed versions are also made.

Care of haymaking machinery
Lubrication of the power drive, bearings and gearboxes must not be neglected. Tyre pressures should be checked regularly. The tines must be kept tight and any which are broken must be replaced.

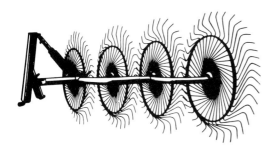

FIGURE 14.3   Finger wheel swath turner.

SUGGESTED STUDENT ACTIVITIES

1. Look for the different types of mower and conditioner and study their construction.
2. Find out how to replace the blades on a disc or drum mower.

PLATE 14.8   *Side raking a crop, previously spread to hasten drying. This type of tedder has flexible rubber rotors with spirally arranged tines. Note the transport wheels lifted above the tedder.*

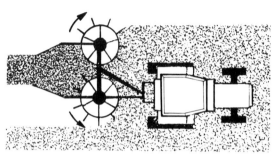

Windrowing a spread crop

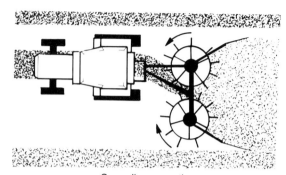

Spreading a swath

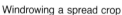

Note the position of the rear doors for each operation

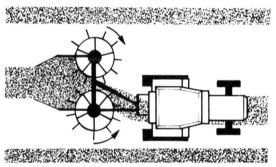

Turning a swath on to fresh ground

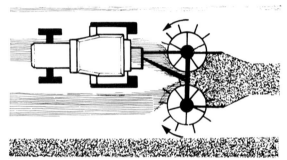

Putting two swaths together

FIGURE 14.4   Rotary tedder settings.   (*Lely*)

3. Look for the different types of haymaking machinery.
4. Find out how many different swath treatments can be carried out with one type of hay machine on a farm you know.

SAFETY CHECK

Never allow anyone to walk behind, or work near, a disc or drum mower when it is cutting. There is a risk of serious injury from flying stones and other debris.

Chapter 15

SILAGE MAKING MACHINERY

Most farmers use a forage harvester to harvest green crops for silage. Forage crops are normally cut with a mower conditioner and, after wilting, the swath is picked up with a forage harvester, chopped into short lengths and blown into a trailer. Maize for silage is harvested with a special maize cutter attachment, then chopped and conveyed to a trailer. Some farmers prefer a forage harvester which cuts, chops and loads the crop in one operation. Small-scale methods of harvesting silage include cutting with a mower, then collecting the swath with either a buckrake or a forage wagon.

FORAGE HARVESTERS

All types of green material, used to make silage, can be harvested with a mounted, trailed or self-propelled forage harvester. Most trailed harvesters have power take-off drive, but some are engine driven. Small forage harvesters are sometimes fitted to the three point linkage, with provision for either side or rear operation. Self-propelled machines are popular with farmers who make large quantities of silage, and with agricultural contractors.

There are two types of forage harvester chop-

PLATE 15.1   *Self-propelled forage harvesters picking up wilted grass for silage making.*   (Claas)

PLATE 15.2   *An off-set flywheel chop forage harvester. The combined chopping flywheel and fan is housed just below the vertical section of the spout.*   (Reco)

ping mechanism—flywheel chop and cylinder chop.

Flywheel Chop Harvesters

This type of harvester has a flywheel chopper with a set of knives and paddle fan blades. The flywheel runs at speeds of 600–1,000 rpm, depending on model. The previously cut crop is lifted by a pick-up cylinder and fed to a cross-conveying auger which transfers it to the chopper unit. Some models of flywheel chop harvester have feed rolls which compress the material to ensure an even feed to the chopping unit. The chopper knives cut the crop against a fixed blade in the opening from the auger to the flywheel housing. After chopping, the fan blades blow the cut material up the delivery spout into a trailer.

There is some variation in the range of chop lengths which can be obtained on different harvesters. One model, for example, can chop the crop into lengths which vary from 4–20 mm.

Some flywheel chop forage harvesters have a flail cutting mechanism instead of a pick-up

cylinder. The flails, which run at a speed of about 1,800 rpm, cut the crop and feed it to the auger conveyor. The flail cutting unit can be replaced with a pick-up cylinder on most harvesters.

Controls and adjustments

Pick-up or cutting height is adjusted hydraulically or with wheels on the pick-up cylinder. Some trailed machines have a height control ram on the drawbar.

Length of chop can be reduced by fitting an extra set of blades on the flywheel. The speed of the feed rolls can be changed, with a slow rate of feed giving a short chop.

Delivery spout   This can be adjusted to deliver the chopped material to a trailer pulled alongside or behind the harvester. The deflector plate at the end of the spout can be adjusted to control the discharge angle.

Drawbar   Off-set, trailed machines require the drawbar to be moved from the road to field position before work can start.

PLATE 15.3 *Cylinder chop forage harvester, off-set on the tractor, with forage trailer towed behind. The box on the chute contains the electronic control for the chute swivel and discharge chute.* (Claas)

Feed roll reverse Blockages in the feed rolls caused mainly by overload can be cleared by operating the reverse drive mechanism.

Cylinder Chop Harvesters

A more precise length of chop can be obtained with a cylinder chopper. The previously cut swath is gathered by a pick-up cylinder and lifted to the auger which transfers it to the feed rollers. The chopper unit, which may be to one side of, or behind, the auger, receives the crop from the feed rollers. The rollers compress the material into a firm wad so that the knives can slice it cleanly against a fixed shearbar. The rollers also control the rate of crop flow through the cutting cylinder to ensure a uniform length of chop.

The chopper cylinder is heavy. It runs at a speed of about 900 rpm. A typical trailed harvester has a 630 mm diameter cylinder, with 8 blades, 420 mm wide. An example of a large

self-propelled harvester has a cylinder of the same diameter with 12 knives, 660 mm wide. The length of chop can be varied, for example, from 4–70 mm. The chopper cylinder also provides the airflow which carries the chopped material up the delivery spout into a trailer. Some high output harvesters have an auxiliary impeller fan which assists in blowing the material into a trailer.

Controls and adjustments

Height of the pick-up is set with either a hydraulic ram or wheels on the pick-up cylinder. Some harvesters have both methods of adjustment.

Delivery spout angle and direction can be set for side or rear delivery. The deflector plate at the end of the spout can be adjusted to control discharge angle. These adjustments may be made either electrically or hydraulically. Older machines may have manual adjustment.

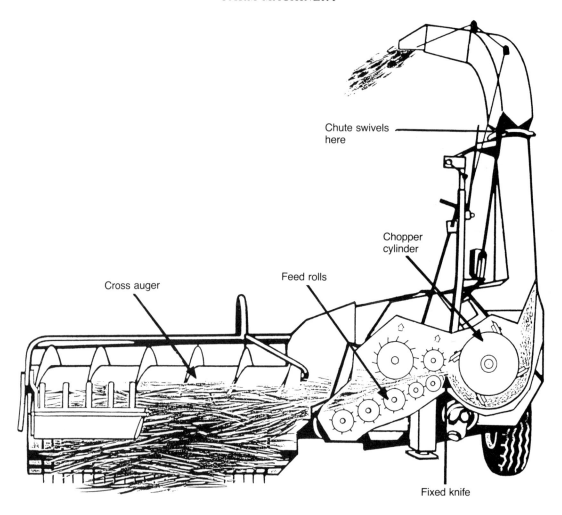

FIGURE 15.1 Sectional diagram of a precision chop forage harvester. (*Claas*)

Length of chop Two adjustments are provided. The speed of the feed rolls can be altered and the number of knives on the cylinder can be changed.

Feed roll reverse is provided on most harvesters. This enables the driver to clear blockages caused mainly by overload.

Safety devices
All forage harvesters have mechanical protection devices which are designed to prevent damage to the machine. They include:

- *Main drive protection* Provided by slip and over-run clutches. The slip clutch allows the power drive to continue turning when there is a blockage in the machine. The over-run clutch allows the machine to slow down at its own pace when the power drive is disengaged.
- *Metal detector* Forage harvesters are liable to pick up pieces of metal which will do serious damage to the chopping unit. The detector locates ferrous metal objects magnetically. The entire feed mechanism is stopped instantly and, by engaging the reverse drive, the metal object can be removed after stopping the machine.
- *Shear bolts* Fitted in the main drive of many forage harvesters.
- *Vee-belt drives* When used they provide a

PLATE 15.4 *Forage harvester chopper cylinder. It has eight banks of three chopping knives. Replacement of a damaged knife is a simple task.* (Claas)

built-in safety mechanism, since overload will cause belt slip.

Knife sharpening

Many flywheel and cylinder chop mechanisms have a built-in knife sharpener. A grinding stone is attached to guide rails which are set parallel to the edge of the knives. The cylinder is turned slowly under power and the stone is drawn from side to side along the guide rails, either mechanically or by hand. The stone can be set closer to the knives as they are gradually ground away. The shearbar must also be sharpened from time to time to ensure a clean cut. The clearance between the shearbar and the knives is critical. The correct setting is achieved on most machines by adjusting the shearbar. On some harvesters the cylinder is adjusted up to the shearbar.

The importance of sharp knives, carefully adjusted to the shearbar, is illustrated by the fact that a ten-knife cylinder running at 1,000 rpm will make 10,000 cuts per minute.

Attachments

Most forage harvesters have a pick-up cylinder which lifts a previously cut swath into the machine. This can be replaced on many models of forage harvester with one or more of the following:

- *Flail cutting unit* A high speed horizontal rotor with swinging flails cuts the crop which is then fed to the auger and chopper unit.
- *Rotary mower* A disc mower unit used for cutting grass and similar crops can be attached to some models of harvester.
- *Maize header* Used to harvest maize for silage, this attachment is in common use. Depending on size of machine, the maize header may have the capacity to cut from one to six rows in one pass. Each row of maize is guided by dividers to a cutting disc and auger unit from which it passes to the chopping unit.
- *Cereal mower* For farmers who use green cereal crops to make silage. The crop is harvested with a cutter bar and reel unit of similar design to that used on a combine harvester.

Power requirement

Forage harvesters have a rather high power requirement. A large self-propelled harvester with the capacity to operate a five-row maize header or a 2.75 m pick-up cylinder will be powered by an engine of around 224 kW (300 hp). A trailed model with a 1.7 m pick-up cylinder needs a tractor with an engine of around 85 kW (115 hp).

Flail Forage Harvesters

This simple design harvester has been replaced on many farms by a flywheel or cylinder chop machine. The flail forage harvester produces long pieces of lacerated grass which are only suitable for pit or clamp silage making. Three or four rows of flails which turn at a speed of about 1,600 rpm cut the grass and then lacerate it against a shearbar across the full width of the rotor. The flails also create an air blast which blows the lacerated grass into a trailer pulled alongside or behind the harvester. The flails are power take-off driven through a gearbox and multiple vee-belts.

The main adjustments are height of cut, set with the harvester wheels and drawbar, delivery spout angle and degree of laceration, which is increased by setting the shearbar nearer to the flails.

Width of cut is from 1–1.5 m. Flail harvesters may be trailed or mounted with the cutting

PLATE 15.5 *Forage harvester with maize harvesting attachment.*   (John Deere)

mechanism either off-set to one side of the tractor or behind it.

Maintenance of forage harvesters

Lubrication points must receive daily attention during the season. Follow the lubrication schedule given in the operator's handbook. Gearbox oil levels and tyre pressures should be checked regularly. Belt and chain drives must be correctly tensioned.

Cutting flails, used on some harvesters, must be sharp and secure. Any missing or badly damaged flails must be replaced immediately to avoid rotor vibration. Flails should always be replaced in opposite pairs to maintain correct rotor balance. Nuts and bolts, especially on components which run at high speed, must be kept tight. Keep all cutting edges sharp: blunt knives give inferior results and waste power.

THE BUCKRAKE

A buckrake consists of a number of strong steel tines about 1.4 m long, bolted to a tine bar which pivots on a frame attached to the three-point linkage. The tines are tipped to drop the load with a lever which releases a locking catch holding the tines in the transport position. The tines return to the transport position when the buckrake is lowered to the ground.

Buckrakes can be attached to a front end loader, sometimes with a hydraulically operated push-off mechanism. This enables the driver to unload the buckrake well above ground level to build up a silage heap.

A buckrake is used by reversing the tractor along the swath with the tines lowered, until it is fully loaded. The tines are pitched (angled) slightly downwards to give a clean pick up. Too much pitch may result in soil being gathered up

PLATE 15.6 *Forage wagon: it picks up the swath, chops it and then loads it in the wagon. A chain and slat floor conveyor unloads the wagon from the rear with the tailgate lifted.* (Claas)

with the swath. The crop is deposited at the clamp by tipping the tines. Economical use of the buckrake is limited to very short hauls from field to clamp, or very small-scale production.

Forage Wagons

Swaths of wilted grass can be collected with a forage wagon, a harvesting system which lends itself to silage making on smaller farms. A pick-up cylinder at the front of a high sided wagon lifts the crop to feeder tines and a chopping mechanism. The chopped material passes into the wagon until it is fully loaded and then hauled to the silage clamp or pit. The load is discharged from the back of the wagon with a moving floor conveyor after opening the rear door, usually with a hydraulic ram. Some forage wagons have a cross-conveyor at the rear which can be used for direct feeding. Forage wagon capacities vary

from 10–40 cubic metres. A typical 30 cubic metre wagon has a power requirement of 29 kW (40 hp); it gives a chop length of 40–240 mm, depending on the number of knives used.

Silage Handling and Storage

Silage may be made in pits dug in the ground, surface clamps, silage towers or plastic bags.

Pits and clamps
Rough terrain fork trucks, front end loaders or buckrakes can be used to fill silage pits or build clamps. Loading and consolidating silage, especially when building an above-ground clamp, has certain dangers. Never drive close to the edge when consolidating the clamp. Wide track settings will increase the stability of the tractor. Dual rear wheels are a useful aid to safe and effective consolidation of silage.

Towers

Although expensive to install, towers make complete mechanisation of silage making and feeding possible. Reinforced concrete or coated steel sheets are used to construct tower silos. Silos with diameters of up to 10 m are in common use, with heights ranging from 9–24 m.

A dump box or forage box, combined with a blower, is generally used to fill the tower. The dump box is a large hopper on wheels with a floor conveyor to carry the material to a discharge chute at the back of the hopper. A forage box is a self-emptying trailer with high sides; it can be towed by a tractor alongside a forage harvester. The blower is a high output fan, driven by a tractor or an electric motor which fills the tower by blowing the crop up a duct into the filling hatch.

Grass is tipped into the dump box from trailers and then fed to the blower. Forage boxes discharge the crop into the blower. Some blowers have a chopping unit to cut the material into short lengths before blowing. In any case, green material stored in tower silos must be chopped into short lengths. Long material would be extremely difficult to remove from the tower.

After filling, the tower must be sealed to keep out the air. Provision is made to prevent a build-up of pressure on hot days with an expansion valve or breather bag. Air is excluded from the tower to remove oxygen which is undesirable in the silage making process. Never enter a silage tower as soon as it has been opened; it must be left open to allow oxygen in before anyone is permitted to go inside the tower.

The green material is chopped before it is blown into the tower to make mechanical unloading possible. This can be with a top or bottom unloader. Top unloaders have a series of digging teeth on an endless chain which cut and carry the silage to a blower unit at the centre of the unloader. A chute from the blower throws the silage through an opening in the side of the tower into a duct. From here, it falls to ground level where a conveyor can take the silage to the livestock. The cutting mechanism moves round on top of the silage; it has its pivot and electric motor drive at the centre of the unloader. As the level of silage falls, the unloader is lowered. It is supported from the top of the tower by chains attached to the roof.

A bottom unloader also has a cutting mechanism and conveyor. It is placed in an opening at the base of the tower. The silage gradually moves downwards as the tower is emptied. The unloader can be withdrawn from the tower for maintenance or repairs.

Plastic bags

Silage can also be made by sealing the forage in plastic bags. Round bales, and sometimes rectangular bales, are placed in plastic bags, sealed and left for the fermentation process. Plastic bag silage is popular for small-scale silage making, because handling is convenient and no special storage facilities are needed.

Big bales can also be wrapped mechanically with plastic. A bale wrapper can cover a big round bale with plastic sheet in little more than one minute. After placing a bale on the tractor-mounted wrapper, it is rotated on rollers and covered with plastic from a roll of wrapping film on the machine. On completion of the cycle, the sheeting is cut, the end tucked in and the wrapped bale is tipped off the rollers. The sheeting is well overlapped to ensure an airtight seal.

SUGGESTED STUDENT ACTIVITIES

1. Look for the different types of forage harvester and study their main design features.
2. Locate the adjustments on a forage harvester.
3. Find out how the knives are sharpened on a cylinder chopping unit.
4. Study the working principles and drive layout of a forage wagon.
5. When visiting farms in your area where silage is made, find out which system is used and why it is preferred.

SAFETY CHECK

When using a tractor to consolidate a silage clamp keep away from the edges. Use dual wheels if possible. Ensure that the clamp is made properly with the sides well supported. It should have guard rails or edge markers along each side.

Consolidating a silage clamp is not a suitable job for an inexperienced driver.

Chapter 16

BALERS

A baler picks up loose hay or straw, compresses it into bales of even size and weight, then ties them with twine or wire. The completed bale is discharged from the back of the baler where it falls to the ground. Many farmers collect the bales in some form of bale sledge towed behind the baler. Heaps of bales can then be left at intervals across the field. There are many systems and machines for handling and carting bales from the field to the store.

There are two main types of baler. One makes small rectangular bales, weighing less than 25 kg. The other makes large bales, up to 750 kg, which may be rectangular or cylinder shaped (round bales).

THE PICK-UP BALER

Both big and small balers are pick-up balers, but the term pick-up baler usually refers to the machine which makes small bales. The other type is known as a big baler.

PLATE 16.1 *A big rectangular baler at work.* (Hesston)

PLATE 16.2   *A pick-up baler with a bale accumulator.*   (Claas)

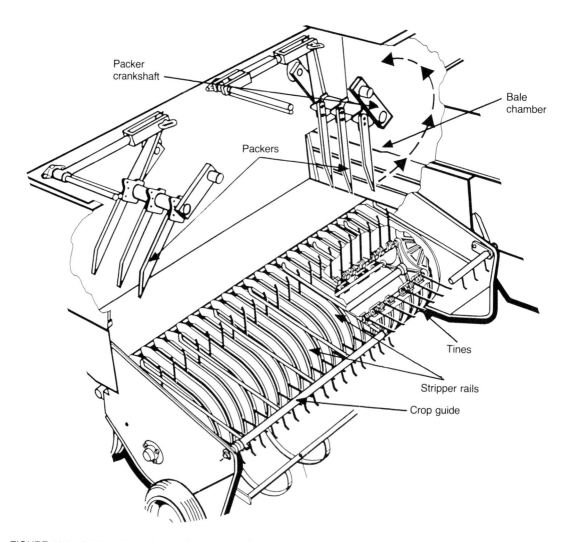

FIGURE 16.1   Detail of the pick-up cylinder and packers.   (*Claas*)

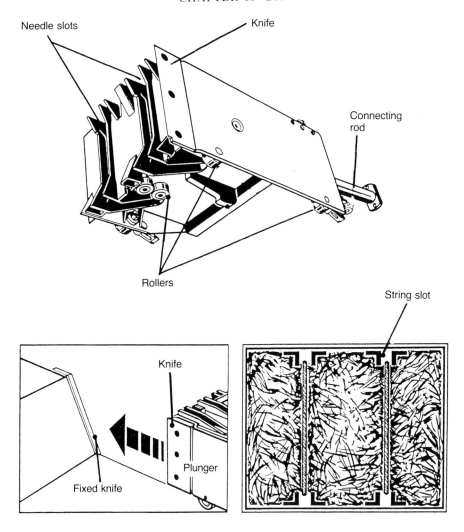

FIGURE 16.2   Detail of plunger and knife.   (*Claas*)

Pick-up balers make bales of around 25 kg. The most common dimensions are 460 mm wide by 360 mm high by 920 mm long. Bale length can be varied from 0.3–1.3 m but for easy stacking the bale length is usually twice its width.

A typical machine consists of a pick-up cylinder which lifts the crop from the ground to an auger or other type of side-conveying mechanism. A spring-loaded crop guide, above the pick-up cylinder, guides the crop to the auger or packers and it also stops the wind blowing the crop about in blustery conditions. The crop guide must be in place at all times as it is also a

safety guard to protect the operator from injury by the pick-up tines.

The packer fingers pass the crop from the auger, when fitted, to the bale chamber. Some models of baler use the packer tines to collect the crop from the pick-up and feed it into the bale chamber in one operation. Once in the bale chamber the crop is compressed by the plunger (or ram) which makes about eighty strokes per minute. It is driven by a heavy crankshaft and connecting rod from a gearbox at the front of the baler.

Each wad of material entering the bale

chamber is trimmed by a knife on the pick-up side of the plunger. It cuts the crop against a knife fixed to the rear of the bale chamber feed opening.

A star-shaped metering wheel measures the length of the bale being made. When it reaches the correct length, the tying mechanism is tripped (engaged) by a connecting link from the metering wheel to the knotter drive trip clutch. When the previous bale was tied, a twine from each knotter was placed through the bale chamber by the needles. These twines encircle the bale as it is formed. The twine is carried in a twine box which holds four spools. It passes through a twine tensioner and a series of guides before reaching the needles and knotters. The twine tensioner regulates the flow of twine.

The tying cycle starts with the needles entering the bale chamber. Both needles place twine around the bale before they enter the knotters. With the knots tied, the needles return to the rest position below the bale chamber, leaving two twines in position ready for the next bale.

Tying the knot

The entire tying cycle is completed in less time than the plunger requires to make one stroke. This is about half a second. There are two knotters.

The three basic parts of the knotter are the bill hook, retainer disc and the knife. These are driven by gears and are timed to operate at a certain point in the tying cycle.

When the needle withdraws from the knotter after tying a knot, it leaves the end of the twine in the retainer disc. It passes down through the bale chamber, through the needle eye and on to the twine box.

When the knotters are tripped the needles enter the bale chamber. Each needle places the twine into a retainer disc which turns slightly to grip the twine. The bale now has twine completely round it, with the ends gripped by the retainer. The twine runs over the top of the bill hook which is just in front of the retainer. The knotter drive gears turn the bill hook which wraps the twine around itself. The bill hook jaws open, take hold of the twine, then snap shut gripping it tight.

The knife is attached to a swinging stripper arm. At this point in the tying cycle, the stripper arm swings across, and pulls the twine off the bill hook in such a way that the knot is formed. The twine is cut above the knot at the same time. Some knotters have a fixed knife and the twine is cut as the stripper arm pulls it across the knife. The needle returns to the rest position and leaves a twine in the retainer ready for the next knot.

Controls and adjustments

Engaging the drive Most balers are power take-off driven and drive is engaged with the power

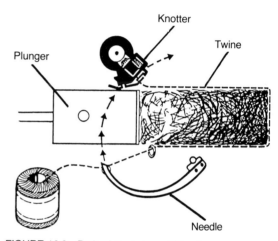

FIGURE 16.3 Path of the twine. (*Claas*)

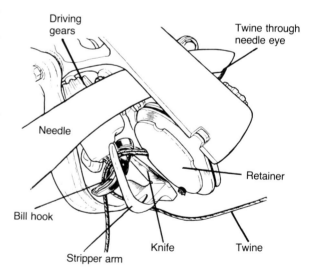

FIGURE 16.4 Baler knotter.

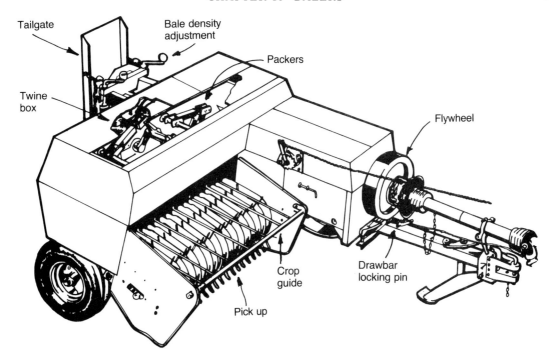

FIGURE 16.5 Pick-up baler controls.

take-off lever. A few balers are engine driven. These have a belt tensioner pulley engaged with a lever.

Pick-up height Set with a hand lever or a small wheel at the side of the pick-up cylinder. The tips of the tines should be about 100 mm above the ground: low enough for a clean pick-up but not too low so that soil or stones are collected. The pick-up cylinder has a balance spring to help it follow uneven field surfaces.

Packer adjustment The packers feed the crop into the bale chamber. They can be adjusted to suit different crops and swaths. The feed from the packers must be even to ensure a well-shaped bale. When they place too much material to one side of the bale chamber, curved or banana shaped bales will be made. This fault is corrected by changing the throw (movement) of the packer fingers. In a light crop, the packers will need to travel further into the bale chamber than they do for a heavy crop. This adjustment is usually made by resetting the packer drive linkage.

Bale density The weight of the bale is altered with the bale density adjusting screws at the end of the bale chamber. Both screws must be at the same setting or curved bales are made. The adjusting screws increase the resistance against the bales as they move along the bale chamber. When the adjusters are tightened, resistance increases and the bales are heavier.

Another way to increase bale density, when the adjusting screws do not produce a heavy enough bale, is to fit bale chamber wedges. These are bolted to the sides of the bale chamber in pairs, one opposite the other. Their effect is greatest when bolted at the end of the bale chamber near the tailgate.

Bale length An adjustable stop on the metering wheel linkage is used to alter bale length.

Drawbar The baler pick-up cylinder must be off-set when in work to allow the tractor wheels to run on cleared land. The drawbar lock is released and then the baler can be moved to the road or field position according to need. Never engage the power take-off drive when the drawbar is in the road position. This can cause serious damage to the drive shaft and bearings.

Line of drive

The power take-off drive is transmitted through a heavy flywheel to a gearbox. Here a pair of bevel gears put the drive through a 90 degree angle. A crankshaft on the gearbox output shaft gives the plunger its backwards and forwards movement. Drive to the pick-up, cross conveyor, packers and tying mechanism is also taken from the main gearbox. Except for a vee-belt-driven auger on some balers, all drives are by chain and sprocket. Chain drives are used because the packers and needles are timed in relation to the plunger.

The packers must enter the bale chamber as soon as the plunger clears the feed opening. The needles enter the bale chamber through protective slots in the plunger as it moves toward the rear of the baler. Always make sure the packer to plunger or the needles to plunger setting is correct before connecting up the drive chains after carrying out a repair or replacing a broken chain. Check the timing instructions given in the baler handbook.

Safety mechanisms

Balers have a number of protective mechanisms to reduce the risk of damage through overloading, blockages or mistiming. These are in addition to the guards to protect the operator from injury. The guards must be in place whenever the baler is in use.

The main drive Power from the tractor to the baler flywheel is transmitted by a single shear bolt in the flywheel hub. A sudden heavy overload will break the shear bolt, preventing more serious damage. Always use the correct shear bolt for your baler.

A friction slip clutch on the flywheel absorbs the less violent overloading. An over-run clutch is usually included in the main drive. This acts like a bicycle freewheel, allowing the baler to slow down at its own pace after the power take-off drive has been disengaged.

The pick-up cylinder has either a slip clutch or an over-run clutch to protect it. The slip clutch has two serrated faces held together by spring pressure. When the pick-up is overloaded, or if

the tines catch the ground, the clutch will slip and make a loud clattering noise.

The over-run clutch prevents the pick-up turning backwards if the outfit is reversed with the pick-up cylinder fully lowered.

The packer fingers are protected by an overload spring, and on some models with a shear bolt in the packer drive. Some balers have front packer fingers made of wood or soft metal. The plunger knife will cut through the soft material if they accidentally come into contact with it. This is most likely to happen if the packer drive is mistimed.

Needles and knotters The main risk is damage to the needles. The plunger will cut the tips off the needles if they enter the bale chamber at the wrong time whether it is due to wear in the needle drive linkage or slight mistiming.

To prevent the plunger hitting the needle tips, the plunger stop (ram stop)—a heavy steel lug—enters the bale chamber with the needles. The plunger hits the stop, the main drive shear bolt breaks and the needles remain undamaged. Some balers have a form of plunger stop which is hit by the plunger drive crank. A metal lug protrudes into the path of the drive crank when the needles enter the bale chamber. If something is wrong, the crank hits the stop, and again the shear bolt breaks.

A shear bolt is often included in the knotter drive and a knotter brake holds the knotter shaft in the rest position between tying cycles. A needle brake holds the needles in the rest position. These brakes hold the needles and knotters between the tying cycles. Slight wear in the drive linkages and chains could allow the needles to move a short way into the bale chamber in the path of the plunger.

The auger, when fitted, will have a slip clutch if chain driven. Vee-belt-driven augers usually rely on the belt to slip if there is excessive overloading of hay or straw in the baler.

Maintenance of pick-up balers

Balers must be greased daily during the season. Fast moving parts like the crank and plunger bearings may need twice-daily lubrication. Most of the bearings on modern balers are sealed with their supply of lubricant and require no attention during their working life.

Gearbox oil levels should be checked weekly. Tyre pressures must also be checked at frequent intervals. The wheel on the pick-up side of the baler usually has a lower pressure than the other wheel. Check the settings in the baler handbook.

Drive chains and belts must be at the correct tension. Chain drives should be lightly oiled. The plunger has adjustable runners or rollers. The adjusters must be used to keep the plunger knife running close to the bale chamber knife. The bales will have one very ragged edge if the knife clearance is too wide or if it is blunt. Keep the knife sharp and maintain a clearance of about 1 mm between the knives.

At the end of the season clean the baler thoroughly and replace any worn or damaged parts. Coat bright parts with rust preventative liquid and lubricate all round. Store the baler under cover with blocks under the axles, if possible, to take the weight off the wheels.

Using pick-up balers
Where convenient, a cleaner pick-up of the swath can be achieved by using the baler in the opposite direction to the combine or the side rake. An even feed to the bale chamber is essential for well-shaped bales. This depends mainly on the quality of the swath.

Hay bales are much heavier than straw bales and will need less tension on the bale density screws. When the swath is a little damp the density adjusters must not be over-tightened or the bales may heat.

Tying faults may be due to knotter troubles, but first check the quality of the string and make sure the knotters are correctly threaded. Check that the string guides are in good condition and not restricting the flow of twine.

Remember that curved bales may be due to uneven bale density adjustment or faulty packer setting. Bales with one ragged edge occur when the plunger knife is blunt or the knife clearance is too great. Inconsistent bale length is usually caused by uneven feed to the bale chamber.

BIG BALERS

An important advantage of the big baler is the low number of bales produced per hectare. This makes bale handling from field to store a simple task, provided that suitable handling equipment is available.

There are two types of big baler: one makes large round bales, the other large rectangular-shaped bales. The dimensions of big bales vary from one model to another. Round bales can vary from 900 mm to 1.8 m in diameter and be from 1.2–1.5 m wide. Big rectangular bales vary in size from 800 mm to 1.6 m wide by 600 mm to 1.3 m deep and from 1–2.5 m long.

The weight of big bales also varies considerably, depending on size and material baled. A typical round baler has a range of bale weights from 250–350 kg in straw and 350–500 kg in hay. Similar bale weights can be achieved with big rectangular balers.

Round Balers

Both tractor and baler wheels straddle the swath with the pick-up cylinder lifting the crop to the feed rolls which direct it to the bale chamber. There are a number of bale forming mechanisms:

Variable bale chamber The bale is formed by a number of endless belts, which run on rollers. As more material is fed into the baler and the bale size increases, a pivot arm which supports two of the rollers moves upwards to increase the size of the bale chamber. A spring-loaded tensioner maintains constant belt tension. When the bale is made, it is secured with several turns of twine but, before this can be done, forward travel must cease. It is not possible to continue feeding material into the bale chamber during the twine-wrapping cycle. On completion of this process, the rear door of the bale chamber is opened by a hydraulic ram and the bale is ejected. The door is closed and the baler can move forward again to start making the next bale.

Another type of variable bale chamber has an endless slatted chain conveyor with closely spaced slats, attached to drive chains at each side of the machine. As the bale size increases, the chain tension arms pivot, allowing the bale chamber to increase in size until the bale has reached its pre-set diameter. The bale is then secured with twine and ejected.

Fixed chamber with endless belts The bale chamber consists of, in one example, five separate sets of short endless belts supported by rollers across the full width of the chamber. As the bale

PLATE 16.3   *A big round baler making 1.2 m wide bales with a maximum diameter of 1.35 m. The tractor and baler straddle the swath.*   (Ford New Holland)

PLATE 16.4   *A fixed size bale chamber showing the chain and slat conveyor. The pick-up can be clearly seen at the bottom of the chamber. The hydraulic rams used to open the tailgate are fully extended.*   (Krone)

increases in size, the crop comes in to contact with more of the belt sections until it is fully formed. At this point, twine is wrapped and the bale ejected.

Fixed chamber with endless chain conveyor   The pick-up feeds the crop into a fixed size bale chamber. The bale is formed by an endless slatted chain conveyor running round inside the closed bale chamber. When the bale is made, twine is wrapped and the bale ejected.

Fixed chamber with rollers   The pick-up feeds the crop to a circular bale chamber formed by a number of dimpled steel rollers, each separately driven by a heavy duty chain. Again, on completion of the bale, twine is wrapped and the bale ejected.

Wrapping the bales
Round bales can either be wrapped with twine or net. When the bale is formed, the tractor driver

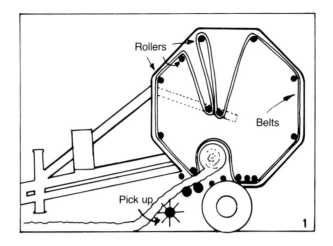

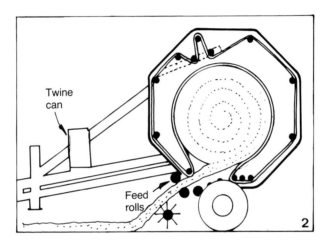

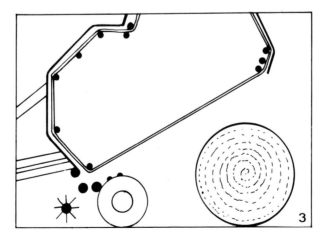

FIGURE 16.6   Diagram of variable chamber layout.

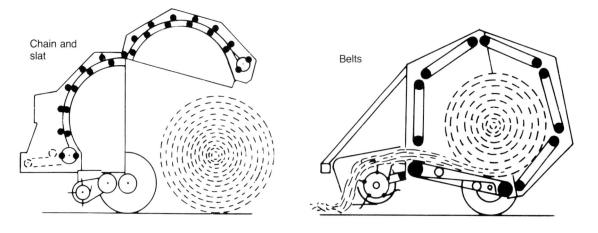

Chain and slat

Belts

FIGURE 16.7   Diagram of fixed bale chamber layouts.

PLATE 16.5   *Fixed chamber round baler with steel rolls making net wrapped bales.*   (Claas)

PLATE 16.6 *Net wrapping attachment fitted on top of a big round baler. This automatically wraps each bale with full width netting. It is a faster process than twine wrapping.* (Claas)

receives a signal that he must stop forward travel. A control is provided which starts the process of automatically wrapping the twine spirally around the bale. When sufficient twine has been used, the end is cut and tucked into the straw. Many balers have a control which allows the driver to select the number of twine wrappings required for the bales. They will be more secure if several wraps are used but this will obviously increase twine costs.

Net can be used as an alternative to twine to hold the bale together before it is ejected from the bale chamber. A net wrapping attachment with a large roll of net automatically wraps the bale when the drive mechanism is engaged. The net is stretched as it leaves the roll to ensure it is tightly wrapped. When sufficient net has been placed round the bale, usually 1½ to 2 turns, it is cut and the bale ejected.

Non-stop Round Balers

It is necessary to stop forward travel with most round balers before wrapping the twine. Some models, however, can be used non-stop, giving a higher output and less wear and tear on the tractor. Non-stop round baling is achieved by means of a pre-bale chamber which stores the crop for the short period while the completed bale is wrapped and ejected. During the twine-wrapping cycle drive to the feed roller, which passes the crop from the pick-up to the bale chamber, is disengaged. After the bale has been ejected the drive is automatically re-engaged and the next bale is started.

Another type of non-stop round baler has two separate bale chambers. When the bale in one of the bale chambers has been completed and is ready for the wrapping cycle, a computerised

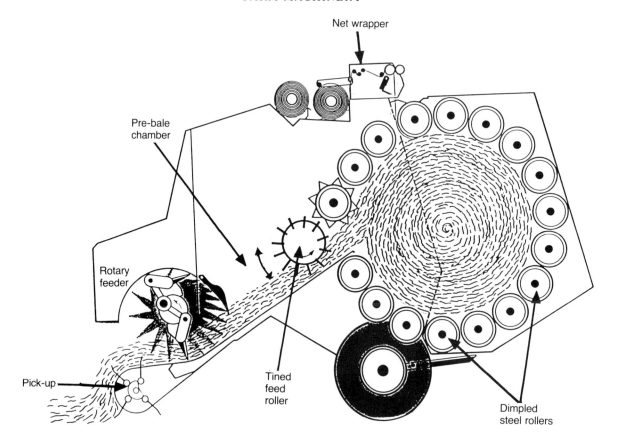

FIGURE 16.8   Non-stop round baler with steel roller-type fixed bale chamber.   (*Claas*)

control diverts the crop to the second bale chamber. The bale in the first bale chamber is then wrapped and ejected while baling continues.

Non-stop round balers are ideal machines for large farms and agricultural contractors who need to cover large areas as quickly as possible.

Baler control systems

Command and control systems employing computer technology can be used to make the control and monitoring of round balers a simple matter. A control panel in the tractor cab, linked to the baler, tells the driver exactly what is happening in the bale formation and wrapping cycles. The unit has a memory which, once set, will continue to repeat the same baling cycle to give bale diameter and number of wraps required. The unit provides an over-ride function, which enables the driver to wrap a partly made bale

when the field is completed, and remote control for opening and closing the rear door of the bale chamber.

Big Rectangular Balers

Some big rectangular balers are very large, power hungry machines needing tractors of up to 100 kW (140 hp) to operate them. Typical bale weights for these large balers are 900 kg in hay and 650 kg in straw. Outputs can exceed 40 hectares (100 acres) per day, making this type of baler very much a contractor's machine. An example of a smaller model of big rectangular baler, suitable for a large arable farm, needing a 70 kW (100 hp) tractor, will make straw bales weighing about 240 kg.

Both tractor and baler straddle the swath, which needs to be quite large to provide suf-

PLATE 16.7 *Big rectangular baler.* (Massey-Ferguson)

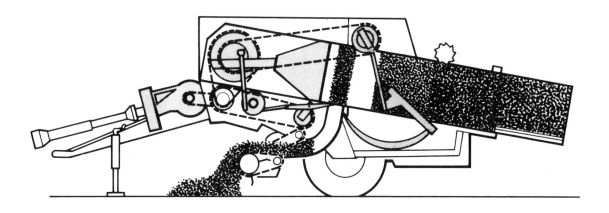

FIGURE 16.9 The crop is collected by the pick-up and fed to the packer tines in the curved feed chamber. It is formed into wads and pushed up to the bale chamber by the crop as it leaves the pick-up. The ram runs at about fifty strokes per minute to form the bales. The star wheel on top of the bale chamber trips the tying cycle when the bale reaches the required size. (*Massey-Ferguson*)

PLATE 16.8   *A mechanical bale accumulator which leaves groups of bales arranged in 'flat eights'. The bales are loaded on to a trailer with a bale grab on a tractor loader or fork truck.*   (Brown)

PLATE 16.9   *An accumulator for big bales. The bales are formed into threes by hydraulic rams on the accumulator floor and then tipped off. An indicator panel in the tractor cab shows the operating sequence. It handles bales 790 mm square.*   (Brown)

PLATE 16.10 *Forty bales can be carried at one time with this squeeze transporter. It is reversed up to a stack of bales which is gripped by the sides and carried to the stack. A double-acting hydraulic ram squeezes the sides of the transporter against the bales.* (Brown)

ficient material to keep the feed mechanism full. The pick-up lifts the swath to the charge chamber where the individual wads which form the complete bale are made. Each wad is fed into the main bale chamber and compressed by a heavy plunger, operating at about fifty strokes per minute, against the material already in the chamber. Four, five or six knotters, depending on bale cross-section and model, are used to tie twines around the bale when it reaches the required length. After tying, the bale moves along the bale chamber and drops onto the ground.

Some big balers use wire as an alternative to string. One such baler has four wire twisters which secure four bands of wire around the bale. The extra strength provided by the wire ties makes it possible to produce very high density bales.

In-cab control and monitoring systems are widely used with big rectangular balers. Bale density can be set and maintained automatically to suit changing crop conditions. Such functions as knotter performance, plunger operation and bale shape can be monitored and adjustments made where required. Plunger strokes per minute, power take-off speed, bale counting, pick-up speed, etc. can also be checked.

Using big balers

Careful swath preparation is important: swaths must have a regular shape and even density. It will often be necessary to combine two or more swaths when using a big baler behind a combine harvester with a narrow cutter bar. The tractor wheels must be set wide enough to straddle the crop.

Silage making can be carried out with a big

PLATE 16.11   *A big bale gripper, similar to the flat eight version, used to load big square bales.*   (Brown)

round baler. The wilted crop is baled, carefully stacked in a clamp and covered to exclude the air. Large plastic bags can also be used to seal individual bales ready for the fermentation process.

BALE HANDLING EQUIPMENT

Various types of equipment can be used to handle bales, ranging from a simple sledge which collects the bales as they leave the baler to expensive self-propelled bale collectors which load and transport the bales to the stack.

Bale collectors and loaders
Small bales are usually collected by a sledge towed behind the baler. It is usual for the bales to be left in particular formations, such as a 'flat eight' where two rows of four bales, set together, are left ready for loading with a special attachment on a tractor loader or rough terrain fork lift truck. Sledges which leave flat eights and similar bale formations may be operated hydraulically or, more usually, by mechanical means.

Bale wagons which pick up, load and transport small bales provide a fast, one-man bale collection system. The load of bales is tipped and stacked with hydraulic rams. Bale wagons can be towed by a tractor or be self-propelled.

Big bales are handled singly, one of their great advantages being the fact that the labour-intensive task of collecting many small bales is eliminated.

Transporting bales
Large flat floored trailers are used to haul both big and small bales. Loading and unloading may

PLATE 16.12   *A tractor loader with a parallel lift linkage handling big round bales. The linkage keeps the bales parallel to the ground or stack, no matter what height they are above the ground.*   (Massey-Ferguson)

be with a tractor loader or a fork lift truck fitted with a suitable loading attachment.

For short hauls, individual big bales may be transported from the field with an attachment such as a spike or gripper on a tractor loader or fork lift truck. Tractor-mounted transporters, usually carried on the three point linkage, with a capacity of about forty bales, are suitable for short haulage work.

Big round bales can be transported around the farm and taken to cattle yards with a tractor-mounted carrier which has two heavy spikes. The spikes are pushed into the centre of a round bale by a hydraulic ram. When the carrier is lowered, the round bale can be unrolled like a carpet, after removing the twine, by pulling it slowly forward with the tractor. This provides an easy method of bedding down livestock housed in yards.

Bale Wrappers

These machines are used to wrap bales of green material for silage with an airtight plastic film. Bale wrappers may either be trailed or mounted. The bale to be wrapped is placed on two rollers which are mounted on a turntable either with a squeeze loader which is an integral part of the wrapper or with a separate tractor loader. The wrapping cycle is started by attaching the stretch film plastic to the bale. A combination of rotating and turning the bale results in it being covered with a complete bandage of plastic film. The number of layers applied and the amount of overlap of each layer of film can be adjusted.

Once wrapping is completed, the bales can be left in the field until it is convenient to haul them to the farmyard. The bales can be transported

PLATE 16.13   *Tractor-mounted big bale wrapper. The machine is reversed up to a bale and plastic film is tied to the twine. The bale is rolled along the ground by slowly reversing the tractor. A hydraulically driven arm, carrying a roll of stretch plastic film, rotates around the bale, wrapping it with overlapping layers until it is completely covered. The plastic film is cut and tucked in to secure it in position.*   (Wilder)

PLATE 16.14   *Turntable type big bale wrapper. The bale to be wrapped is placed on the turntable which rotates it in two directions. Powered belts turn the bale while the turntable rotates it horizontally. Plastic film from the roll is attached to the bale; it is then wrapped with overlapping layers of film. The platform is tipped to remove the bale after wrapping. This machine can be mounted on a tractor or be free-standing. It may be driven by the tractor hydraulic system or by its own hydraulic power pack.*   (Wilder)

with a spike loader but the holes must be patched. A squeeze grab, used with care, should lift and transport the bale without damage to the plastic.

SUGGESTED STUDENT ACTIVITIES

1. Locate the field adjustments and mechanical safety devices on a pick-up baler.
2. Learn how to thread the needles on a baler. Study a knotter and identify the main components.
3. Look at round balers at an agricultural show. How many different types of bale chamber system can you find?

4. Watch out for the different types of bale handling systems in use. Note how each system works.
5. Look for a bale wrapping machine, and study its working principles.

SAFETY CHECK

Never attempt to clear a blockage in a pick-up baler without first stopping the tractor engine. Disengaging the power take-off drive is *not* enough, because someone could accidentally re-engage the drive. The consequences could be serious.

Chapter 17

COMBINE HARVESTERS

The combine harvester is used for a wide range of crops, including cereals, grass and clover seed, peas, beans, oil seed rape, linseed and maize. It cuts the crop, conveys it to the threshing mechanism, then sorts the grain from the straw and chaff. The straw is left in a swath on the ground behind the machine, the chaff is blown out of the back. The grain is conveyed to a grain tank (hopper) from which it is unloaded with a high capacity auger into a trailer.

The size and type of farm will determine choice of combine. Mixed farms will normally have a harvester with a 3.2–4.9 m cutter bar. Large arable farms may have a combine with a cutting width of up to 6.7 m with a capacity of up to 20 tonnes per hour.

It is necessary to remove wide cutter bars before the combine can be moved along the road. A special cutter bar trailer towed behind either the combine or a tractor is used for this task.

The combine harvester is a self-propelled machine, with the largest models powered by an engine of up to 175 kW (240 hp) or more.

PLATE 17.1 *A combine harvester at work.* (Massey-Ferguson)

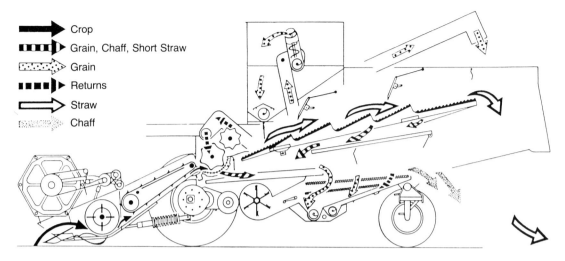

Crop
Grain, Chaff, Short Straw
Grain
Returns
Straw
Chaff

FIGURE 17.1 Crop flow through a combine harvester.

Hydrostatic transmissions provide a stepless range of forward and reverse speeds, including a fast road speed. Modern combines have an air-conditioned cab, hydrostatic steering, finger-tip controls within reach of the seat and a monitoring system to warn the driver if the threshing performance is below standard.

CROP FLOW

There are three stages in the flow of the crop through a combine harvester:

- Cutting and elevating.
- Threshing.
- Separation.

Cutting and Elevating

The crop is cut with a reciprocating knife cutter bar which makes about 1,000 cuts per minute. The knife sections may have either serrated or smooth cutting edges. Serrated knife sections require no sharpening; they are ideal for cutting clean cereal crops. When there is green material, or the crop is undersown, a smooth sectioned knife may be required.

The reciprocating knife is supported by fingers bolted to the cutter bar frame. A ledger plate on each finger provides the second edge of the scissor-type cutting action. Knife clips hold the sections close to the cutting edges of the fingers to give a clean cut.

There is a divider at each side of the cutter bar platform. It separates the crop to be cut from that which will be cut on the next round. Grain lifters can be fitted to the cutter bar to help lift laid or tangled crops. They give a much improved cutting performance in difficult conditions.

The crop is cut while it is held against the knife by the reel. After cutting, the reel directs the crop on to the cutter bar table or platform. A pick-up reel is standard equipment; it has spring steel tines attached to the reel slats which can be angled to improve cutting efficiency. Reel height and speed can be adjusted, usually hydraulically, from the driving cab.

Most combines cut a swath about three times as wide as the main elevator and threshing cylinder. An auger with both left- and right-handed flights moves the crop to the middle of the platform. Retracting tines at the centre of the auger feed the crop to the main elevator. Normally of chain and slat construction, the main elevator lifts the crop to the threshing cylinder. A stone trap collects solid objects brought on to the cutter bar platform. This helps to prevent stones and other hard objects damaging the threshing mechanism. The stone trap must be emptied regularly; a trapdoor is provided for this purpose.

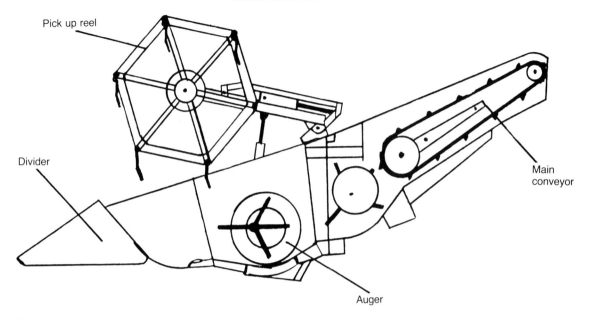

FIGURE 17.2   The cutter bar platform and elevator.

Threshing

The crop is fed from the elevator to the cylinder (drum) and concave. Feed is sometimes assisted by a rotary front beater which speeds up the crop as it passes from the elevator to the cylinder and concave.

The threshing mechanism consists of a heavy rotating cylinder with rasp-like beater bars. The concave is a stationary set of bars below and very close to the cylinder. There are spaces between the concave bars for the threshed grain to fall through onto the grain pan (or tray). The grain is rubbed from the ears as the crop passes between the cylinder and concave. In good combining conditions, at least three-quarters of the grain should fall through the concave together with some chaff and short lengths of straw. When harvesting cereals, the cylinder speed will be about 1,800 m/min. (This is equivalent to 6,000 ft/min which is about 60 mph.) The cylinder speed for peas is about half of that needed to thresh cereal crops. Cylinder diameter varies with different models of combine so threshing speeds are given in m/min. This is the peripheral speed of the cylinder and refers to the distance travelled, in this case by the rasp bars, in one minute.

Peripheral speed can be calculated by multi-plying the cylinder diameter (m) by the cylinder rpm and then by 3.142. The result of this calculation is the peripheral speed of the cylinder in metres per minute. Combine harvester instruction manuals give the threshing speeds for different crops and conditions in rpm. A typical combine with a cylinder diameter of 560 mm will run at about 1,000 rpm when combining cereals. This can be converted to a peripheral speed of 1,800 m/min. The distance between the cylinder and concave must also be adjusted to suit different crops.

The rear beater is the final part of the threshing unit. It slows the crop as it leaves the cylinder and directs it down on to the straw walkers or other system of separation.

The Separating Mechanism

There are two parts to this:

- Separation of the remaining grain from the straw.
- Separation of the good grain from the partly threshed heads, small grains and weed seeds.

One of the following systems is used to separate the remaining grain from the straw.

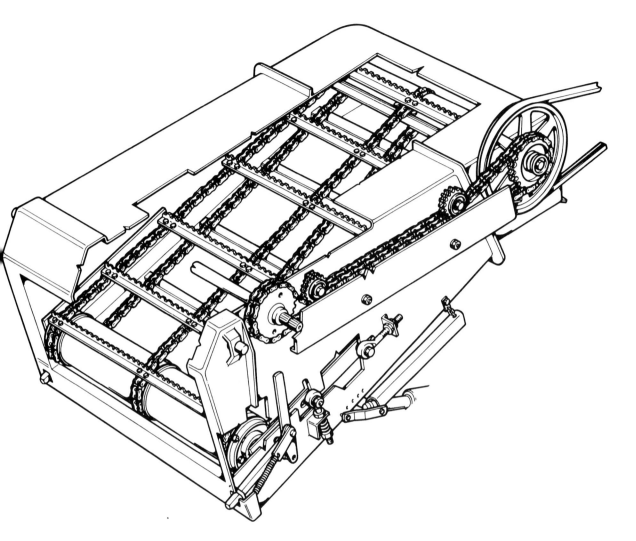

FIGURE 17.3   The main elevator showing detail of the chain and slat conveyor which lifts the crop from the table to the threshing cylinder.   (*Fiatagri*)

Straw walkers

The crop passes to the straw walkers which convey it to the back of the machine. Loose grain is shaken from the straw by the rising and falling action of the walkers, which are driven by a pair of crankshafts. The grain is directed to a tray underneath the concave. Some combines have a trough under each straw walker which directs the loose grain back to the tray below the concave. One or more canvas curtains hang from the roof above the straw walkers. They slow down the passage of the straw as it moves along the straw walkers, to ensure as much loose grain as possible is removed. Extra curtains may be required when combining peas and beans or when handling heavy cereal crops.

Cylinder separation

After leaving the threshing cylinder and concave, the straw passes between a series of separating cylinders and concaves. The layer of straw moves from one separating cylinder to another until it reaches the back of the combine. Any loose grain in the straw falls through the separating concaves

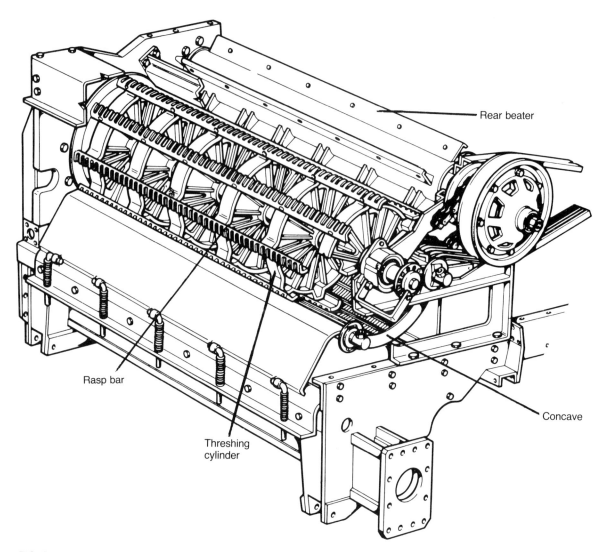

FIGURE 17.4   Threshing cylinder and concave.   (*Fiatagri*)

and falls to the grain sieve where it joins the grain from the main concave.

Rotary separation

Two additional rotors and concaves are used with a standard straw walker system. As the crop leaves the threshing cylinder, it passes to a rotary beater running at about two-thirds of the drum speed. The beater, which has its own concave, removes more grain from the straw before it passes to the rotary separator which

runs at a lower speed than the beater. The rotary separator consists of a rotor and concave. The rotor has a number of beater bars which have rows of large teeth. The separator removes almost all of the remaining grain from the straw before it passes to the straw walkers. The grain removed by the beater and the separator finds its way to the main grain pan and sieves.

This system of rotary separation is also used with one model of combine harvester which has a twin rotor system of straw removal, instead

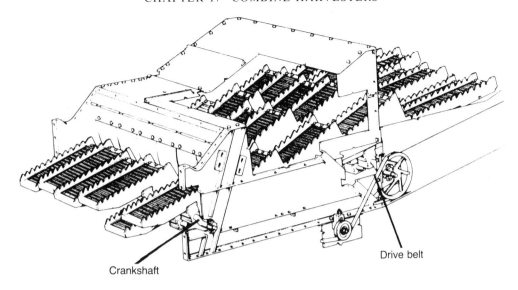

Crankshaft

Drive belt

FIGURE 17.5   Combine harvester straw walker assembly.   (*Fiatagri*)

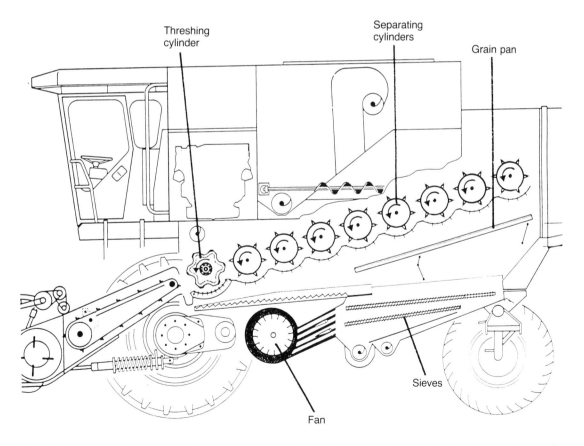

Threshing
cylinder

Separating
cylinders

Grain pan

Fan

Sieves

FIGURE 17.6   Sectional view of a combine harvester with cylinder straw separation instead of straw walkers.   (*Claas*)

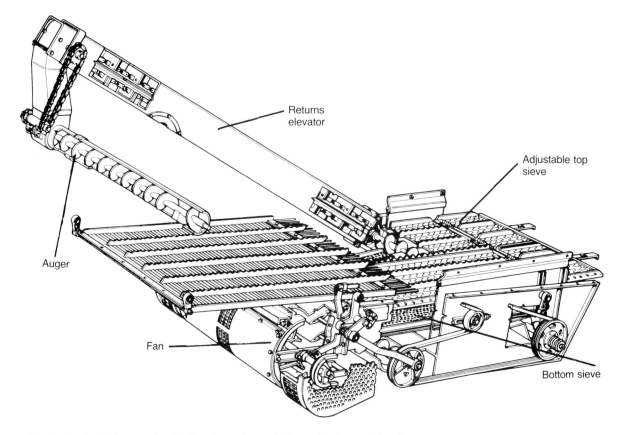

FIGURE 17.7   Grain separator showing sieves, fan and returns elevator.   (*Fiatagri*)

of conventional straw walkers. Any remaining grain is removed centrifugally as the straw passes to the back of the machine.

Grain removed from the straw by this system passes to the sieves where it joins the bulk of grain removed by the cylinder and concave.

Sorting the grain

The grain and chaff on the grain pan below the concave must now be sorted. It has arrived there from the concave and the straw walkers or other separation mechanism. The grain separating unit consists of two sieves, one above the other, and a fan. The shaking action of the sieves separates the good grain from the partly threshed heads, broken straw and chaff. A fan provides a constant flow of air to keep the sieves free from blockages.

The top sieve (chaffer sieve) allows everything through except long straws and partly threshed heads. Most combines have an extension to the

top sieve with larger holes in it. These allow the partly threshed material to fall through to the returns auger.

The bottom sieve (grain sieve) receives the material from the top sieve. It separates the good grain from the rest of the material which falls on to it. Some combines have a grain pan under the top sieve which directs the material to the front of the bottom sieve. The good grain and any smaller grains and weed seeds pass through the bottom sieve. These are collected by the grain auger and carried to the grain elevator which lifts it to the grain tank.

A second auger—the returns auger—collects any partly threshed heads which pass over the back of the top and bottom sieves and carries them to the returns elevator. The returns are lifted to the cylinder for rethreshing. Some combines offer the choice of directing the returns to the cylinder or to the top sieve depending on the condition of the returns.

The harvested grain is conveyed to a tank or hopper situated at the top of the machine. It is usual for an auger system to be installed in the tank to ensure even filling. Grain tank capacities vary according to the size of the harvester. For example, a typical 6 m cut machine has a tank capacity of 8,000 litres (this will hold about 6.5 tonnes of wheat). Another model with a 3.6 m cutter bar has a tank capacity of 4,200 litres (3.5 tonnes of wheat).

A high capacity unloading auger, which is swung into position over a trailer with a hydraulic ram, can empty the grain tank in about two minutes. Unloading can be carried out on most combines while cutting continues. Many combines have a warning light which flashes when the tank is almost full. This alerts the driver, who has the task of hauling the grain to the store, that he must get to the combine as soon as possible.

Special attachments

Straw chopper   Many combines have a straw chopper attached to the back of the machine. It cuts the straw into very short lengths and spreads it over an area of similar width to the cutter bar. The chopper has rows of high speed swinging knives which cut the straw against a row of fixed knives. When it is not required, the chopper can be disengaged and the straw will be left in a swath.

Straw spreader   This unit spreads the straw over the stubble over a wide area in readiness for the plough. Most farmers prefer to chop and spread their straw if it is not wanted.

Pick-up attachment   Some farmers prefer to cut certain crops such as oil seed rape, grasses and clover with a swather. The swath is left to dry

PLATE 17.2   *A combine harvester chopping and spreading the straw as it leaves the straw walkers.*

(Ford New Holland)

and then combined at a later date. The pick-up attachment is fitted to the cutter bar table. It lifts the swath and feeds it to the elevator which conveys it to the threshing cylinder.

Maize picker This attachment is infrequently used in Great Britain since it is difficult to ripen maize for combining. The picker is similar to that used on a forage harvester when gathering green maize for silage. The cutter bar and reel is removed and the maize header is attached in its place. Four-, five-, six- and eight-row headers are available.

CONTROLS AND ADJUSTMENTS

The combine harvester must be set to suit the particular crop being harvested. There are considerable variations between the settings for grass seed and those for beans. The settings given below are only a guide; the instruction book must be consulted when preparing the combine for work. Once in the field, fine tuning can be made where necessary.

The Cutting Platform

Efficient threshing requires an even feed to the cylinder and concave. Careful adjustment of the various components of the cutting platform will help achieve this.

The dividers
They should be set to part the crop cleanly. Straw must not be allowed to build up around the divider points.

The reel
When working in a standing crop, the following settings are a useful guide.

- The reel slats should contact the straw just below the ears.
- The reel speed should be slightly faster than the forward speed of the combine. As a guide, the reel rpm should be slightly higher than the rpm of the front wheels. The reel speed is adjusted hydraulically from the cab. Older machines require different size sprockets to be fitted to the chain drive when a change in reel speed is needed.
- The reel slats should be set forward of the

knife. The reel is moved backwards and forwards, either with a hydraulic ram or mechanically.

In laid or tangled crops:

- Lower the reel.
- Increase reel speed.
- Angle the tines towards the knife. In this position they can lift the crop to give a cleaner cut.

The knife
The cutter bar and knife must be in good condition:

- The knife sections must be undamaged and securely riveted to the knife back.
- When a smooth sectioned knife is used, it must be kept sharp.
- Adjust the knife clips to ensure that the knife sections are held firmly against the cutting edges of the fingers.
- Check the knife is correctly registered. It should travel from the centre of one finger to the centre of the next or next but one according to the knife drive system.

Cutting height
This is set with a lever in the cab, which controls the hydraulic rams which lift and lower the table. The usual cutting height is 100 mm from the ground. Many combines have an automatic height control system which maintains a pre-set stubble length. It is a useful aid on uneven fields and reduces the risk of picking up stones.

In heavy crops there is a risk of feeding too much material into the combine, especially with smaller output models. This may result in grain being lost from the back of the combine with the straw.

When it is necessary to reduce the quantity of straw taken into the combine it should be done by selecting a slightly slower forward speed or by raising the cutter bar a little. Never reduce the crop intake by driving in a way which uses less than the full width of the cutter bar. This will result in an uneven feed to the cylinder, giving poor quality threshing.

The table auger
This must be set to avoid blockages or straw wrapping around the flights. The retracting tines

should be set to gather the crop and feed it to the elevator. The clearance from the end of the tines to the platform can be adjusted to suit crop conditions. The tines must not touch the bottom of the platform. The auger flights must be set at a suitable height above the platform for the crop being harvested. The flights will rub grain from the ears or pods if set too low, especially in heavy crops.

The stone trap
This should be emptied daily to ensure solid objects do not reach the cylinder and concave.

The Threshing Unit

Careful adjustment of the cylinder and concave is important to ensure efficient threshing. Settings will depend on the type of crop, its condition and its moisture content.

The cylinder
The speed should be as low as possible to give complete threshing. It is adjusted, on all but the oldest combines, by a pair of hydraulically operated variable speed vee-belt pulleys. A typical cylinder speed for threshing cereal crops is 1,800 metres per minute. For example, a 560 mm diameter cylinder achieves this speed when running at 1,000 rpm.

Low cylinder speeds will not remove all the grain from the ears. High speeds will crack the grain and, in dry conditions, break up the straw which will find its way into the grain tank, spoiling the sample.

The concave
The clearance between the cylinder and the concave should be set as wide as possible to give complete threshing. A wide clearance leaves unthreshed grain in the ears. Grain will be damaged if the concave is set too close to the cylinder. The concave clearance is always greater at the front than it is at the back.

A typical clearance for wheat is 10 mm at the front and 4 mm at the back. Adjustment is made with a lever in the cab.

The relationship between front and rear concave clearance is retained automatically on most combines. Some models, however, have separate levers for setting front and rear clearances. Even with these machines, only the front clearance lever is used for normal adjustment; the ratio of front to rear clearance is retained. The second lever can be used to alter the rear clearance; it will also alter the ratio between front and rear settings.

Filler plates are placed between the concave bars when threshing small seeds such as grass and clover to improve threshing performance. No seeds can pass through the concave: separation takes place on the straw walkers. The concave is set very close to the cylinder when combining grass and other small seeds.

When combining barley, filler plates can be fitted between the front concave bars to help de-awn the grain.

The straw walkers
No adjustments are required, but it is important to keep them clean. Blocked walkers will prevent grain reaching the sieves, and it will then be lost out the back with the straw. Drive belts must be kept at the correct tension. A slack belt will slip, reduce shaking speed and result in grain loss.

The sieves
Both top and bottom sieves can be adjusted on most combines. The size of the sieve opening is altered with a lever. Some machines have a round hole bottom sieve which must be changed to suit different crops.

Adjustable sieves should be set as wide open as possible to obtain a good sample. Losses will occur if the sieve is closed too much, and the sample will be poor if the sieve is set too wide.

The top or chaffer sieve is set with an opening of about 10 mm for cereals. Settings for other crops will depend upon the size of the seed.

The bottom or grain sieve will be set at approximately 7 mm for cereals. Some farmers prefer a round hole bottom sieve. Cereals need a sieve with 16 mm diameter holes and grass requires a 2.5–4.5 mm sieve. The use of a 5 mm round hole sieve is recommended for combining rape. This will give a much improved sample by preventing most of the pieces of pod and stalk which fall through the top sieve reaching the tank.

The disadvantage of round hole sieves is the need to have a number of them to deal with a variety of crops. However, they tend to produce a cleaner sample because short pieces of straw do not easily pass through the round holes.

PLATE 17.3   *Combine harvester with hillside attachment. This keeps the threshing and separating systems level when working on sloping land. The cutter bar follows the ground contours.*   (Fiatagri)

The fan

Adjustments are provided to control the direction and quantity of air. Deflector plates in the fan outlet can be set to direct the air to the front, centre or rear of the sieves. The quantity of air is controlled by the speed of the fan. Some combines have an adjustable air inlet for the fan which provides an additional control of airflow.

Light seeds need a small amount of wind at the front of the sieves. Cereals need rather more wind, directed to the centre of the sieves, and peas need large quantities of air to keep the sieves clean. As a general guide, use as much wind as possible without blowing seed out of the back of the combine. Too little wind will cause blocked sieves.

Controls for Driving and Combining

There are numerous controls in the combine cab in addition to those which regulate the cutting and threshing mechanisms. The cab has a full set of instruments which indicate threshing and separating efficiency, cylinder speed, identify grain loss, record engine speed, forward speed and hours worked, etc.

Combines with high specifications have computer systems in an air-conditioned cab, which can store the complete instruction book, record output per hour, yield from the field, monitor grain loss, etc. The information can be displayed on a monitor screen or as a print-out for a permanent record.

PLATE 17.4 *Combine harvester cab with driving and threshing controls within easy reach of the seat. The wide expanse of glass gives a full view of the cutter bar and reel.* (Fiatagri)

The driving controls include a lever to select speed ranges and reverse, with a second control to give stepless speed variation within each range. Hydraulically operated variable speed pulleys or hydrostatic transmission provide the stepless speed control. The throttle is used to set the engine at maximum speed when combining and vary it when the machine is out of work. A pedal or switch in the cab can be used to stop the drive to the cutter bar, reel and auger if an object which could damage the combine is picked up. On some combines, a reverse drive for the cutting platform is provided too.

There is a full set of instruments for monitoring the engine, transmission, threshing mechanisms and the tank unloader. Reel speed, cylinder speed, fan speed, cutting height, etc.

are controlled from the cab, either with levers and pedals or electronic switches which operate small motors and rams. For example, the tank unloading auger is moved into position over the trailer by a hydraulic ram which is activated by a switch in the cab.

ROTARY COMBINES

The principle of rotary threshing is not new, having been invented nearly 200 years ago. A rotary or axial flow combine is only different in its threshing and separation mechanisms; the rest of the machine is similar to a conventional combine harvester with cylinder and concave threshing and straw walker separation.

PLATE 17.5   *Axial flow combine harvester with straw chopper and automatic reel speed control.*   (Case IH)

The threshing rotor is about 2.7 m long and 610 or 760 mm in diameter, depending on size of machine. It runs at right angles to the cutter bar. The rotor has an impeller at the intake end which draws a supply of air into the harvester at the same time as it conveys the crop from the main elevator to the threshing mechanism. The rotor speed is altered to suit the crop being harvested, with a variable unit drive combined with a high--low range gearbox. Rotor speeds range from about 250–1,200 rpm. The speed for threshing cereals is approximately 850 rpm.

The rotor threshes the crop by rubbing it against a concave and separating grate as it travels spirally upwards towards the back of the machine. The grain is thrown outwards by the rotor, passing through the separator grate to a conventional grain sieve and fan. The air drawn into the harvester by the impeller removes most of the dust from the rotor. A discharge beater at the end of the threshing rotor removes any remaining grain from the straw as it leaves the back of the combine.

Another type of rotary combine has the threshing rotor directly behind the cutting platform. The feed opening from the auger is off-set, directly in front of a feed roller which conveys the crop to the rotary thresher. This consists of a threshing rotor and a separating cylinder. After the rotor has threshed the crop, any grain left in the straw is removed by the rotating separator (see Fig. 17.9).

The most unusual thing about this design is that the straw is discharged from the machine at this point. It is left in a swath at the side of the combine just behind the cutting platform. A chopper or a spreader can be attached to the straw outlet.

The grain collected from the cylinder and the separator is conveyed to a large auger. This lifts it to a second transverse auger, which distributes the grain evenly to the sieves for final separation. The fan provides two blasts of air, one through the sieves and the second passes through the grain as it falls from the distribution unit. The grain is finally elevated to the tank in the usual way.

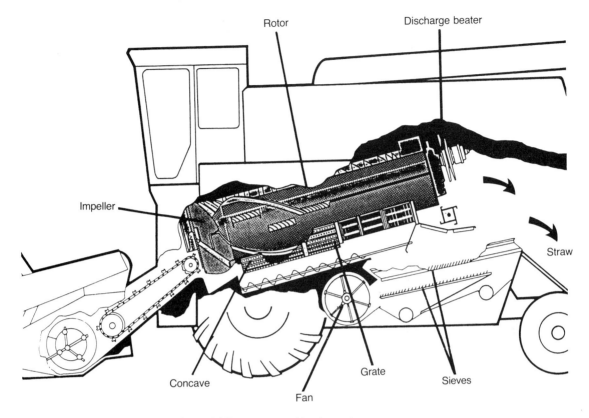

FIGURE 17.8   Sectional diagram of an axial flow rotary combine harvester.

Most of the adjustments on a rotary combine are similar to those for a conventional machine. The important differences are setting the rotor speed and the clearance between the rotor and the concave.

The overall length of a rotary combine is less than that of a straw walker machine with a similar output. The claimed advantages are a higher working speed together with more gentle handling of the crop.

MAINTENANCE OF COMBINES

- Regular cleaning during the season is important. A dirty combine will lose grain out of the back. The concave, straw walkers and sieves must be kept clear of chaff and other rubbish. Combining in damp conditions often results in clogged cylinder bars and a choked concave.
- The stone trap should be emptied daily.

- Correct tension must be maintained on all belt and chain drives.
- The elevators which convey the crop to the cylinder, the grain to the tank and the returns for rethreshing should be checked regularly for damage and correct chain tension.
- The cutter bar should be checked for broken knife sections and damaged fingers. The knife must be a close fit on the fingers and, when a plain section knife is used, it must be sharp.
- Thorough and regular lubrication of all grease nipples is vital. Bearings which are lubricated and sealed for their working life require no attention.
- Check the oil levels in the engine, transmission system, hydraulics and other gearboxes at regular intervals. Top up when necessary.
- The engine has to work in very dusty conditions, which makes daily servicing of the air cleaner an essential part of the maintenance schedule. A blocked air cleaner will restrict airflow and reduce power output.

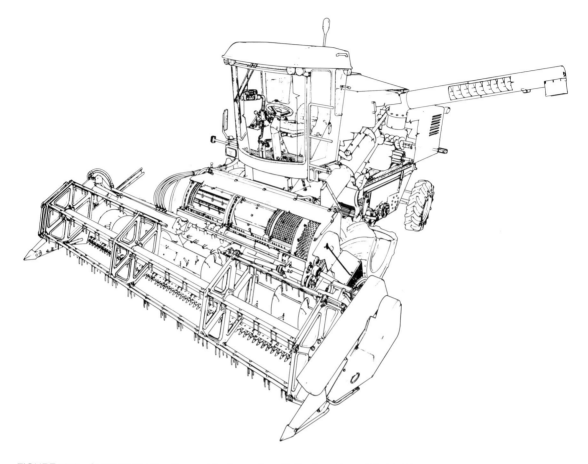

FIGURE 17.9   Sectional diagram of a combine harvester with the threshing unit behind the cutter bar table. Straw is discharged from an outlet behind the table. Sieves and a fan separate the grain from the chaff.   (*Fiatagri*)

Storage

The combine harvester is a very expensive machine with a short working season. Careful storage during the long period from one harvest to the next will help ensure a long and efficient working life and keep corrosion to a minimum. Points to remember include:

- Clean the combine thoroughly, removing all traces of grain, straw and other rubbish.
- Lubricate thoroughly and check oil levels.
- Protect bright parts with rust preventative.
- Service the engine, check the anti-freeze strength if used, or drain the cooling system. Keep the battery well charged; remember that a discharged battery will freeze. Some farmers prefer to run the combine at regular intervals during the winter; others remove the battery and store it in the workshop. When it is not intended to run the engine until the next harvest, condensation inside the engine can be reduced by sealing the air cleaner intake and engine breathers.
- Check tyre pressures and inflate as necessary.
- Check the machine for damage, and replace any broken or worn parts.
- Store under cover. Support the cutter bar with blocks to keep it clear of the ground and take the weight off the lift rams and springs. Leave elevator doors and the stone trap open to discourage rats and other vermin from taking up winter residence in the combine.

PLATE 17.6   *The grain unloader spout in position for discharging grain.*   (Claas)

USING COMBINES

The combine harvester presents few driving problems in the field; the skill is in setting the cutting and threshing mechanisms.

The headland is cut first: three to five rounds should be sufficient to provide ample turning space. When opening up a field where the crop on the headland is a little damp, the extreme edge can be left standing until later in the day when it has had a chance to dry. Care must be taken on the first round of the headland not to damage the unloader auger. It is usual to cut round the outside of the field with the auger spout over the crop when there is a risk of damaging it on overhanging trees.

Once the headland has been harvested, the field may either be cut round and round or in lands (sections). The method will depend on the size of the field, its shape and the availability of trailers for grain carting.

Drive the combine to ensure an even feed to the threshing cylinder by using the full width of the cutter bar. In heavy crops, reduce forward speed or increase the stubble height if the straw walkers are being overloaded. Never reduce cutting width, because this is wasteful and can cause damage to the grain during threshing.

Safety Devices

Some combine harvesters have a performance monitor in the cab which warns of faults in the engine and main drives.

Slip clutches protect the main drive systems from damage through overloading or blockages. These include the drive to the reel, table auger, main elevator, grain and returns elevators. The reel overload clutch should be set so that it will slip if the pick-up tines touch the ground.

The vee-belt drive to the straw walkers and the sieves should be tensioned to allow belt slip if severe overloading occurs.

Threshing Problems

Grain lost from the straw walkers or sieves and damaged grain are the main problem areas.

The in-cab monitor provides information about the location and degree of grain loss whenever the machine is in work. The monitor is connected to special sensors which measure the operating speed of major components such as the cylinder and fan. Other sensors detect and measure the amount of grain in the straw and chaff as it leaves the straw walkers and sieves. The driver needs only to press a button to obtain this information and, after making minor adjustments, the results can be checked.

Grain found with the straw behind the combine could be lost in front, underneath or behind the machine. To investigate the source of loss, run the combine for a short distance, stop the machine and carry out the following checks:

1. Look in front of the combine. Loose grain on the ground in the standing crop suggests it is shedding from the ears because the crop is over-ripe.
2. If there is no shedding, look on the cutting platform. Loose grains here may be due to:
   (a) Reel speed too fast.
   (b) Reel set too high so that it knocks grain from the ears.
   (c) The auger flights are too close to the platform floor. This fault will also cause grain to be rubbed from the ears in very ripe crops.
3. Assuming there are no grain losses at the front of the combine, look underneath for signs of leakage from an ill-fitting elevator door or a small hole in the underside of the machine.
4. The final checks are made behind the combine. Shake some straw from the swath over a sheet to see if there are any broken ears or loose grain. The presence of either indicates:
   (a) A wide concave clearance or a low cylinder speed, leaving unthreshed heads in the straw.
   (b) Blocked concave bars, causing loose grain with the straw.
   (c) The straw walkers are blocked or the drive belt is slack.
   (d) The canvas curtain above the straw walkers is missing or out of position.

(e) The straw walkers are overloaded because too much straw is entering the machine. Slow down or raise the cutter bar slightly.

If there is no grain with the straw but loose grains are found in the stubble underneath the swath, the loss will be from the sieves. Likely causes include:
(a) Sieves closed too much.
(b) Blocked sieves.
(c) Insufficient air from the fan or the deflectors are angled too far towards the back of the sieves.

5. Damaged grain in the tank must also be considered a loss because it spoils the quality of the sample. Cracked grains in the tank are due to:
   (a) Threshing cylinder speed too high.
   (b) Concave set too close to the cylinder.
   (c) Failure to keep the machine fully loaded. The straw cushions the grain as it is threshed. If there is too little straw in the combine, grain may be damaged. This problem can also occur when starting work and at the end of a bout.
   (d) Excessive returns—grain and partly threshed ears which pass over the top sieve for rethreshing—can also cause damaged grain. If the top sieve is not blocked, then it should be opened a little to allow more material to pass through to the bottom sieve.

GRAIN STRIPPER

This is an alternative to the standard cutter bar, auger and reel. The stripper removes the ears from the standing corn and feeds them into the combine for threshing in the normal way. The straw is left standing to be dealt with at a later time. The options are to chop it with a tractor-drawn chopper or, providing there is not an excessive quantity, the standing straw can be ploughed in. Baling can, if required, be done with a tractor, mower and baler combination in one pass. Another alternative is to cut the straw after the ears have been stripped. One use for straw harvested in this way is thatching, because it has not been damaged in the same way as

PLATE 17.7 *Grain stripper head attached to a combine.* (Shelbourne Reynolds)

the straw in the swath behind a conventional combine harvester.

The stripper rotor has eight rows of keyhole-shaped teeth which run in an anti-clockwise direction. It strips the ears and flagleaves from the standing crop and throws them backwards to a conventional crop-collection auger and then they are elevated to the threshing cylinder. The stripper header has a canopy over it and is attached to the combine in the same way as a normal cutting platform.

SWATHER WINDROWER

Some crops, such as oil seed rape, are often cut and left in a windrow to ripen before combining. The swather windrower is a self-propelled

PLATE 17.8 *Part of the canopy has been removed to show the keyhole shaped stripper teeth.*

(Shelbourne Reynolds)

PLATE 17.9 *Rape swather.*

machine which cuts the crop with a conventional reciprocating knife cutter bar and pick-up reel. Vertical knives at each side of the cutter bar reduce losses in tangled crops by dividing the material about to be cut from that left for the next bout. After cutting, the crop is moved towards the centre of the platform by two endless canvas conveyors. The centre of the platform has no floor. When the crop reaches the end of the conveyors, it falls through the opening to the ground, leaving a harvested swath for further treatment.

Typical working widths are 3.0, 3.6 and 4.2 m. Power requirement is about 50 kW (70 hp) and the transmission system is usually hydrostatic.

Suggested Student Activities

1. Identify the system of straw separation used on different models of combine harvester.
2. Study a combine harvester instruction book. Pay particular attention to the section dealing with settings and adjustments.
3. Locate the main combine harvester driving and harvesting controls.
4. Look in the straw behind a combine to see if there is any grain with or under the straw. Try to discover the reason for any grain you find.
5. Inspect the grain in a combine harvester tank for condition and damage. Try to identify the reason for any faults.

Safety Check

Do not let anybody ride on the access steps to a combine harvester cab. Children often play in harvest fields: make sure they are in a safe position before moving off with a combine harvester. Accidents only happen to other people's children, but you are other people to the rest of us!

Chapter 18

ROOT CROP MACHINERY

Sugar Beet Harvesters

A comprehensive range of trailed and self-propelled harvesters, which can lift from one to eight rows at a time, is available to the sugar beet grower.

Types of Harvester

Trailed single-row machines
The most basic model is a single-row trailed harvester, driven by the power take-off. It tops one row ahead of the lifter shares or wheels. Once out of the ground, the beet are cleaned by a rod link conveyor as they are elevated to a hopper or tank. When it is full, the tank is emptied into a trailer by a second rod link elevator.

A tanker model can be used by one person, if necessary, with the driver taking the harvester to the clamp for unloading when the tank is full. If there is some distance to travel to the clamp, the hopper can be emptied into a trailer and, when it is full, the driver hauls the loaded trailer to the clamp. Earlier versions of the single-row trailed harvester have a side delivery elevator which puts the roots into a trailer pulled alongside. This harvesting system requires two tractors.

One type of trailed harvester lifts the roots by their tops. Two endless rubber belts grip the tops and pull the sugar beet from the ground.

PLATE 18.1 *Six-row self-propelled sugar beet harvester at work.* (Matrot)

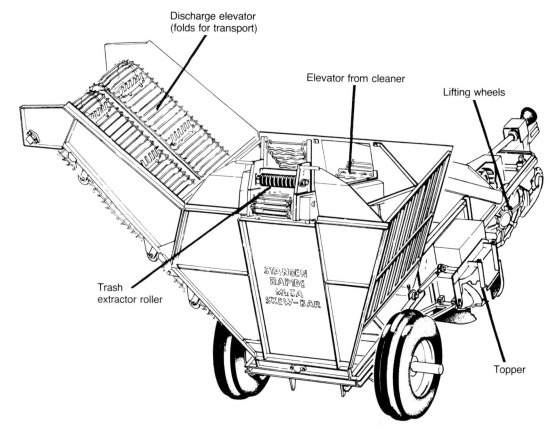

FIGURE 18.1   A single-row tanker harvester. This machine tops one row ahead of the lifter unit. (*Standen*)

The roots are loosened by a share before lifting. The belts convey the sugar beet to two high speed rotating discs which slice off the tops. The roots fall to the cleaning elevator which conveys them to the hopper.

Multi-stage harvesting

This system of harvesting uses either two or three separate machines, each with its own tractor. Each machine carries out one or more operations in the process of harvesting several rows of sugar beet at a time. Multi-stage harvesting is a high speed operation, but much expensive equipment is required. On many farms, where large areas of sugar beet are grown, two- and three-stage systems have been superseded by multi-row, self-propelled harvesters.

Two-stage harvesting consists of a multi-row topper, followed by a second machine which lifts, cleans and loads the beet into a holding tank. More often, the lifter cleaner has a side delivery elevator which puts the beet into a trailer pulled alongside. The three-stage unit employs a multi-row topper, followed by a lifter which deposits the topped beet in a windrow on the ground. The third machine is a cleaner loader which collects the beet, cleans them and elevates them to a bulk trailer.

Several tractors and trailers are also needed, making this harvesting system more suited to a group of growers who contribute towards the cost of the equipment. They also provide the tractors and drivers necessary to keep the machinery at work for very long hours during the harvesting season.

Self-propelled harvesters

There are three main types. One is a complete machine, with versions for lifting from two to eight rows at a time with its own power unit,

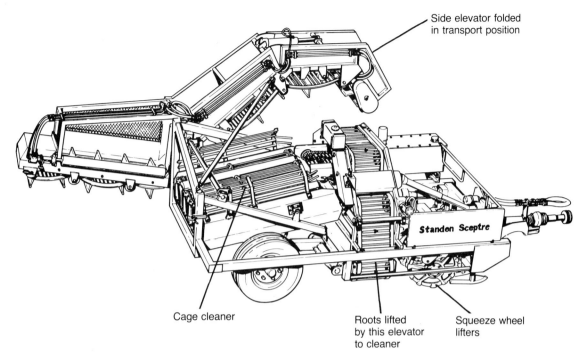

Side elevator folded
in transport position

Standen Sceptre

Cage cleaner

Roots lifted
by this elevator
to cleaner

Squeeze wheel
lifters

FIGURE 18.2   A three-row lifter which cleans and loads sugar beet, previously topped, into a trailer. Provided the tractor is powerful enough, a front-mounted topper can be used at the same time.   (*Standen*)

transmission system and cab. The biggest self-propelled harvester lifts eight rows in one pass and has a 225 kW (300 hp) engine and a large capacity holding tank.

Most single-row, and some two-row self-propelled harvesters are made so that a tractor with an engine power of 55–75 kW (75–100 hp) can be fitted on to the harvester chassis. The tractor wheels are removed before it is installed on the harvester. At the end of the season, the tractor can be taken off the harvester and used for other duties. This task needs heavy lifting equipment and takes about a day to complete.

The third type of self-propelled harvester is built around a high powered tractor. The topper is attached to the front with the lifting, cleaning and loading equipment at the back. Hydraulic motors provide the driving power for most of the harvester components.

Top saving
Some harvesters have a top saver. This provides an efficient mechanical collecting system for farmers who feed sugar beet tops to cattle. An

elevator collects the tops as they are removed by the knife. They are conveyed to a hopper which is emptied at regular intervals across the field, leaving them in windrows ready for collection when required. Multi-row toppers may, as an alternative, have a side delivery elevator or auger which saves the tops by leaving them in large windrows at one side of the machine. Toppers used as part of a two- or three-stage system can have a blower unit at one side of the machine. It receives the tops from the knives or scalpers and blows them into a trailer pulled alongside.

Trailed harvesters, which normally top one row ahead of the lifter, will top two rows ahead when fitted with a top saver. This is necessary to make space for the top saving unit.

Topping

The tops are removed by a fixed knife, guided by a feeler wheel, or by scalper knives used in conjunction with a pre-topping disc or rotor.

The feeler wheel topping unit is carried on a

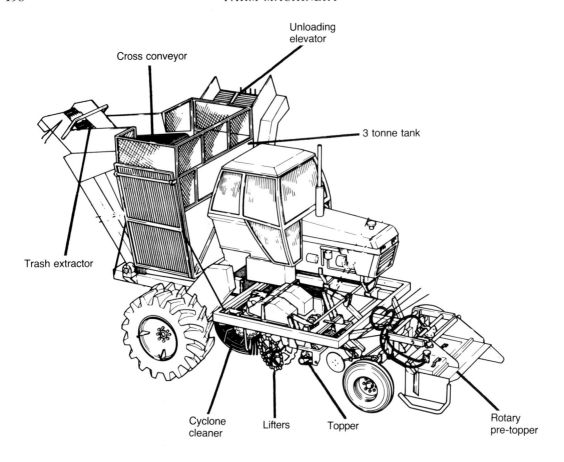

FIGURE 18.3   A two-row self-propelled harvester. The power unit can be used as a tractor after the harvesting season is complete.   (*Standen*)

frame which pivots on the side of the harvester. The feeler wheel and knife are fitted to the frame. A tension spring takes most of the weight of the topper unit, allowing it to float up and down to suit the height of the sugar beet. The feeler wheel runs along the row over the tops to control the position of the knife in relation to the crown of the sugar beet. When a plant is higher than its neighbour, the feeler wheel lifts the knife so that the beet is topped correctly. The feeler wheel is usually chain driven from one of the harvester wheels. A disc is often fitted in front of the topper arm to clear away any trash and reduce blockages.

A spring-loaded device on the knife arm allows the knife to move away from the feeler wheel if it becomes blocked with a large stone or similar obstruction.

After topping, rotating rubber flails sweep the tops off the row to leave a clear space for the lifters. The flails will be positioned either behind the topper unit or in front of the lifter.

Some harvesters have high speed rotary pre-topping knives which remove the bulk of the green material before the feeler wheel and knife complete the topping process.

An alternative system of topping, used on some multi-row machines, consists of a vertical rotor over each row. This cuts off most of the top and is followed by scalping blades which complete the operation. The rotors are usually driven by hydraulic motors. This type of topper may be an integral part of the harvester or a separate machine mounted on the front of a tractor. The tops are moved to one side of the harvester and left on the ground where the roots

PLATE 18.2   *A six-row self-propelled harvester with rotary toppers.*   (Standen)

PLATE 18.3   *Feeler wheel topper showing feeler wheel, knife and trash disc.*   (Agrolux)

PLATE 18.4   *Scalper type topper showing: (A) guide fingers; (B) knife; (C) tines for raking tops away.*   (Agrolux)

have already been lifted. The scalper blades are mounted on a parallel linkage which, with the aid of guide bars, enables them to follow· the height of the roots and trim off any remaining leaf and stem.

Lifting

Squeeze wheels are the most common type of lifter. The wheels run at an angle to each other so that the rims are close together when they are in the ground and further apart at the top. As the wheel rims move together on entering the soil, they grip the roots and lift them from the ground. High speed rotating rubber flails, or small cage wheels, behind the squeeze wheels knock the lifted roots on to a rod link elevator.

Shares are used on some harvesters to lift the roots. Fitted in pairs, the triangular shaped shares squeeze the roots from the ground in a similar way to lifting wheels. After lifting, the roots pass on to a rod link elevator in the same way as squeeze wheel lifted sugar beet.

The shares are bolted to rigid support arms on

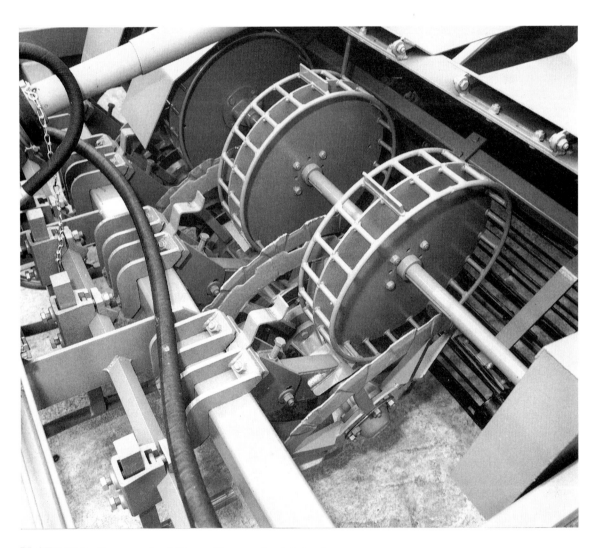

PLATE 18.5   *Squeeze wheel lifters with cage wheels behind which knock the lifted roots on to the elevator web.*   (Standen)

some harvesters, while others have an eccentric cam arrangement which vibrates the shares as they move through the soil.

Cleaning

The roots are cleaned on some harvesters as they are conveyed by the rod link elevator to the holding tank or side delivery elevator. Many harvesters have a second moving chain above the elevator; others have a stationary chain hung above it. In both cases, the upper chain tends to roll the beet and remove most of the soil. The upper chain can be moved closer to the elevator to increase the cleaning action.

Two or more rotary spider wheel cleaners are used on some harvesters. The spider wheels receive the beet from the lifters and remove most of the soil as they are carried to the rod link elevator used to lift the beet to the tank or side elevator.

Further cleaning is achieved by the unloader elevator as it empties the contents of the tank into a trailer. The tank often has a floor and side made of steel bars which allow soil to fall through.

Controls and Adjustments

Topping

The topper must be positioned so that the feeler wheel runs over the centre of the row of sugar beet. It will have to be repositioned when changing from one row width to another. This will only be necessary when moving to another farm where the crop has been drilled at a different row width.

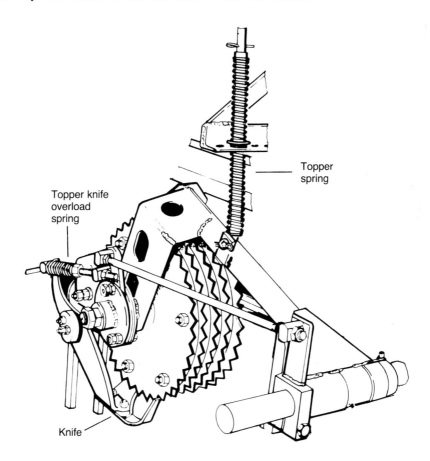

Topper spring

Topper knife overload spring

Knife

FIGURE 18.4 Feeler wheel topper unit adjustments.

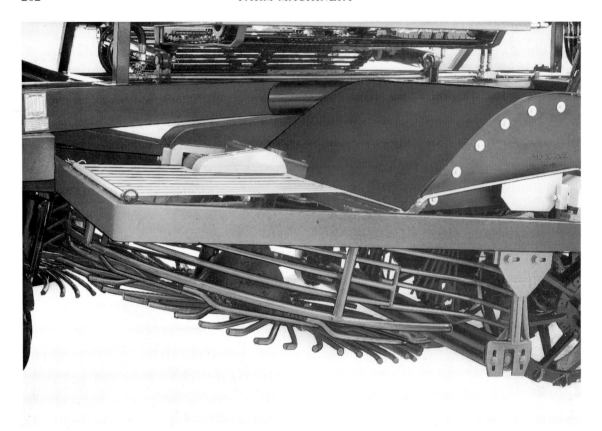

PLATE 18.6 *Spider wheel cleaner.* (Agrolux)

The position of the knife in relation to the centre of the feeler wheel is important; both horizontal and vertical adjustments are provided. The knife should start cutting off the top when the feeler wheel is over the centre of the root. Incorrect setting gives a root with a slanted top. The knife should be moved backwards or forwards depending on which way the top of the root slants. The clearance between the knife and the bottom of the feeler wheel must be set to remove the correct amount of top. If the knife is too far from the feeler wheel, the root will be over-topped; if too close, leaf will be left on the root.

The topper spring, which helps the unit to float up and down according to the height of the roots, can be adjusted. More tension will be needed in light crops where there is a small amount of top. Too little tension on the spring will result in the topper pushing the roots over before they are topped. The knife must be sharp, so it is a good idea to have some spares in the toolbox.

The flails which remove the sliced-off tops have adjustments for height and positioning centrally over the row. Flail height should be set to clear all green material from the row without knocking high standing roots over.

Lifting

Depth of the lifting wheels or shares is controlled with the hydraulic rams which lift and lower them into and out of work. A sensor automatically maintains the pre-set depth of the lifters. Some trailed harvesters have adjustable wheels at the front of the machine which control the depth of the lifters.

The lifters should be set deep enough to remove all of the roots without collecting any more soil than necessary. Overloading of the cleaning mechanism with excess soil will result in reduced output and a dirty sample.

The gap between the bottom of the lifting wheels or the shares can be adjusted to suit the size of the roots. The wheels or shares should be set quite close together at the bottom when harvesting small sugar beet. The gap between the lifting wheels is adjusted either by changing the number of spacing washers between the wheels and their hubs or by means of a lever which moves the rims closer or wider apart.

Cleaning

The problems of cleaning the lifted roots are made worse by setting the lifters too deep. Some harvesters have agitators under the rod link elevators which shake the links to knock soil from the roots. Extra agitators or larger ones can be fitted to give a more vigorous agitation when working in difficult soil conditions. It is also possible to move the stationary cleaning chain closer to the elevator or add additional weights to increase the cleaning performance.

There are two adjustments provided to control the degree of cleaning achieved with spider wheel cleaners. Adjustable gates above the spider wheels can be moved closer to the tines if the roots are not clean enough. It is also possible to change the speed of the spider wheels on some harvesters.

Many farmers use a cleaner loader to remove more soil from the roots when they are loaded into a lorry. A cleaner loader consists of a rod link elevator driven by a small petrol engine. A tractor with a front loader or a rough terrain fork lift truck with a root bucket is used to fill the loader hopper. As the roots are elevated to the lorry most of the remaining soil is removed. In good harvesting conditions the sugar beet will be much cleaner, so they are usually loaded directly into the lorry. When the load reaches the sugar factory, it is sampled for soil and tops. The grower has money deducted from his payment according to the amount of tare (soil and tops) in the sample.

Wheel width

Both tractor and harvester wheels must be set to suit the row width. For example, sugar beet grown at 500 mm row spacing will need the tractor wheels set at 1.5 m. When moving from one row setting to another, it will be necessary to adjust the topper position in relation to the lifter on harvesters which top one row in front of the lifter. Sugar beet rows may be drilled at various widths. Most harvesters offer a range of row widths: a typical self-propelled harvester can deal with rows planted at spacings of 450–600 mm.

Maintenance

Chain and belt tensions, tyre pressures, regular lubrication and knife sharpening are some of the more important maintenance tasks. Self-propelled models need the normal engine and transmission attention as well. Slip clutches, which protect the main mechanical drives from overload, should be checked at the start of the season to ensure they are not seized.

Using sugar beet harvesters

In common with most harvesting machinery, the headlands are lifted first and the field is then worked in lands (sections). Self-propelled machines can move straight into the field and start work. With trailed models, the first row must be topped before lifting can start, and this will be done on the first round of the headland. On the second round, the lifter can also start work. The same procedure is also necessary when opening up the field after lifting the headlands.

POTATO PLANTING EQUIPMENT

A deep seedbed, suitable for planting potatoes, can be prepared on most soils with a plough and a selection of cultivation machinery. However, soils with a high stone and clod content cause problems when the crop is harvested. These include a high rate of wear and tear on the machinery and a poor quality sample containing many damaged tubers. Growing potatoes using the bed system, in soil with the stones and clods removed by specialist machinery, overcomes many problems when the crop is harvested.

The bed system of potato growing requires the seedbed to be ridged up with a bed former. The large ridges are picked up two at a time and sifted by a stone and clod separator. The separator removes the stones and clods and then returns the sifted soil to the ground behind the machine to form the bed. The stones and clods

PLATE 18.7   *Two-row bedformer with subsoiling tines.*   (Grimme)

are deposited in the furrow bottoms between the ridges and will be pushed into the ground by the tractor wheels when the crop is planted. Stones also help with drainage.

Bedformers

A bedformer consists of two, or four, large ridging bodies attached to a mounted toolbar. It is used to make wide ridges in prepared soil. The overall width of a bed, formed by two ridger bodies, will be 1.5–2 m. The width of the bed depends on the position of the bodies on the toolbar. Subsoiler tines can be attached to the toolbar in front of the ridging bodies. They are used to break up the subsoil below the ridger shares. Markers are provided to ensure an accurate join between each pass.

The field should be levelled and cultivated to the full seedbed depth before the land is ridged.

Bed preparation can be speeded up by using a bedformer in tandem with, for example, a rotary cultivator or power harrow.

The furrows between the ridges are used to deposit the stones and clods removed by the separator. The distance between each pair of ridges can be the same as the ridge spacing, or wider if the soil is very stony or cloddy.

Stone and Clod Separators

These are specialist machines, often owned and operated by a contractor. The separator lifts two ridges made by a bedformer, sifts out stones and clods, then returns the soil to the ground. The stones are cross-elevated and deposited in a furrow formed by two of the ridges.

Diabolo rolls control share depth and help guide the machine along the ridges. It is usual for the rear wheels to be steered hydraulically.

PLATE 18.8 *A two-row mounted bedformer, with markers, mounted in tandem with a power harrow.* (Grimme)

As the two ridges pass on to the web-type conveyor, power driven agitators shake and sift the soil which is returned to the ground. Some models have two web elevators, the soil and stones dropping from the end of the first web on to the front of the second. The cascading effect helps to loosen the material on the webs. Either a stationary clod crushing chain, or a clod scrubbing web above the main conveyor, crushes many of the clods picked up with the stones and soil. When the stones and unbroken clods which did not pass through the web conveyor reach the top, they fall on to a cross-conveyor. This carries them sideways and drops them in the furrows between the ridges or beds. It is possible to put all the stones in one furrow or divide the spoil between two furrows.

There is a choice of main conveyor webs, each with a different spacing (pitch) between the bars. A web with a pitch of 40 mm will let fewer stones through with the soil than a wider 60 mm pitch web. The angle of the main conveyor can be altered with hydraulic rams.

By lifting the back of the elevator unit, the separating web is set at a steeper angle. This adjustment improves the quality of stone and clod removal in difficult conditions. There are various systems for operating a stone and clod separator. Some methods require the planting to be done immediately behind the separator, another makes it possible to de-stone without the need to plant at the same time.

Potato Planters

Most potato planters are fully automatic. They are mounted on the three point linkage and feed mechanisms are driven by a land wheel on

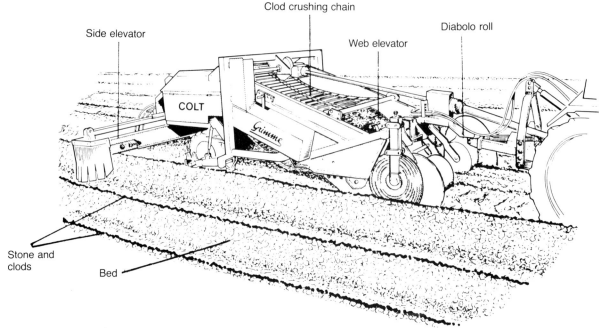

FIGURE 18.5   Stone and clod separator at work.   (*Grimme*)

PLATE 18.9   *Stone and clod separator: the side elevator which places the stones and clods between the beds is driven by a hydraulic motor. The clod crushing chain can be seen suspended above the web elevator.*   (Grimme)

PLATE 18.10   *Two-row potato planter with disc coverers.*

the machine. High output four-row automatic planters are trailed. They have a hopper capacity of nearly 2 tonnes of unchitted seed. A few hand fed planters are still in use.

Potato planters set either two or four rows at a time. A coulter makes a furrow for each row being planted. The hopper, which holds from 500 kg to 1 tonne or more, depending on the model, has an automatic feed mechanism which places the potatoes in the ground at regular intervals. Once planted, ridging bodies or angled discs cover the potatoes with soil.

Farmers who grow potatoes on a small scale may use a hand-fed planter, needing one operator for each row planted. There are two types. One requires the operators to drop a potato down a spout every time a spacer bell rings. Another type of hand-fed planter has a number of cups, each large enough for one potato, attached to either an endless belt or a horizontal rotary conveyor. Drive is by chain and sprocket from the land wheels. The cups are hand filled but tuber spacing is controlled mechanically. Spacing of the potatoes in the row can be altered by changing the speed of feed cups in relation to forward speed. Both chitted (sprouted) and unchitted seed can be planted with hand-fed planters.

One type of fully automatic planter has a cup feed system which collects single potatoes from the hopper and places them at regular intervals in a furrow made by discs or a ridging body.

The planter shown in Figure 18.6 has a belt conveyor in the bottom of the hopper which holds about 500 kg of potatoes. The speed of the belt which carries the potatoes from the hopper to the elevator cups is controlled by a sensor and a magnetic switch. In this way, the sensor can prevent a build-up of potatoes at the base of the feed cup elevator. The potatoes are picked up

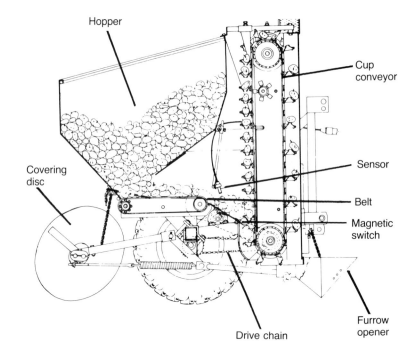

Hopper

Cup conveyor

Covering disc

Sensor

Belt

Magnetic switch

Drive chain

Furrow opener

FIGURE 18.6   How an automatic cup feed potato planter works.

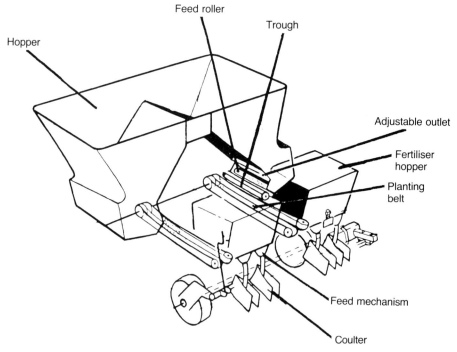

Feed roller

Trough

Hopper

Adjustable outlet

Fertiliser hopper

Planting belt

Feed mechanism

Coulter

FIGURE 18.7   Automatic belt feed potato planter.

by the cups and carried to the outlet point where they fall a short distance into a prepared furrow. Finally, the potatoes are covered with a pair of discs or ridging bodies.

Seed spacing is changed by fitting different sprockets to the chain drive from the land wheel to the cup conveyor. This adjustment provides a choice of tuber spacing in the row from 150–450 mm, while the row width can be varied from 610–910 mm. This machine can be converted for planting chitted seed; it can also have a fertiliser placement unit. It places two bands of fertiliser in the ground through a coulter at each side of and below the row of potatoes. The fertiliser hopper has a capacity of about 200 kg.

A two-row planter for unchitted seed requires a 40 kW tractor. The outfit will plant at speeds of 5–10 km/h (3–6.5 mph).

Another type of automatic planter (see Figure 18.7) has a belt feed system which is less accurate in its spacing of the potatoes but can work at high speed. This planter has a hopper with an adjustable outlet which releases a steady flow of potatoes to the feed roller. The purpose of the feed roller is to fill a shallow trough which holds about ten potatoes. At pre-set intervals, the trough is automatically tipped and the tubers fall on to continuous planting belts. The potatoes are carried to the ground by the planting belts where they are covered by ridging bodies.

The planter has a fertiliser placement unit. Each row has a fertiliser hopper with a metering unit in the outlet. It supplies controlled rates of fertiliser to separate coulters which place it in the ground to each side and below the rows of potatoes.

This machine is intended for unchitted seed. It will plant them at even spacing, provided that the seed has been carefully graded. Spacing is altered by changing the speed of the feed belts, which are driven by a chain and sprockets from the planter land wheels. Fertiliser application is altered by changing the speed of the feed mechanism in the hoppers.

Potato planter adjustments

Row width Row spacings of 610–910 mm are in common use. The furrow openers and coverers can be moved along the toolbar to achieve the required row setting.

Tuber spacing Belt feed machines give rather irregular spacing if the seed is not evenly sized. Tuber spacings of 230–380 mm can be achieved by changing feed belt speed. Cup feed planters are more accurate; the speed of the feed cups can be altered to give a range of row spacings from 150–450 mm.

Planting depth This is controlled by either wheels on the planter or by the tractor hydraulic system. The covering bodies or discs can have their depth adjusted independently, if required, by moving them up or down in their mounting brackets. When this adjustment is made, it may also change the size and shape of the ridge.

Maintenance of potato planters

Regular lubrication of all moving parts, checks on belt and chain tensions and tyre pressures are the main routine maintenance tasks. Fertiliser hoppers should be cleaned out at the end of each day: it may rain next day and fertiliser will corrode metal parts at an alarming rate.

At the end of the season, bright parts should be protected from rust and any parts which have come into contact with fertiliser must be thoroughly cleaned and protected from corrosion.

POTATO HARVESTING MACHINERY

Potatoes can be lifted from the soil with an elevator digger or spinner and left in rows for hand picking. These are relatively cheap machines, used mainly by farmers who grow small areas of potatoes. There are several types of trailed and self-propelled harvester for the large-scale grower.

Elevator diggers

Elevator diggers lift either one or two rows of potatoes at a time. They may be mounted or semi-mounted, usually with a pair of pneumatic tyred castor wheels to support the back of the machine. A wide flat share moves along under the ridge, lifting the potatoes and soil on to a rod link or web-type elevator. Most of the soil is shaken through the elevator to the ground as the crop travels upwards. The potatoes, and any remaining soil, fall to a second, shorter rod link

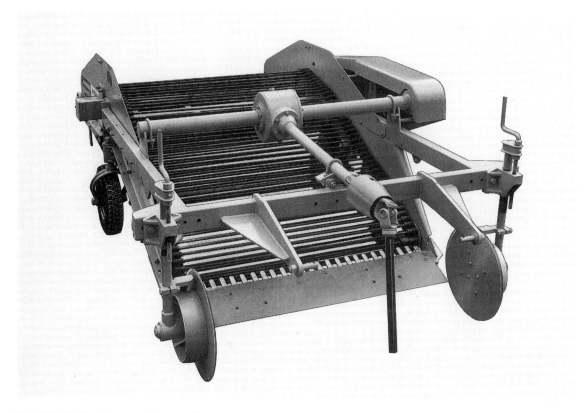

PLATE 18.11   *Two-row elevator digger.*   (Agrolux)

or web elevator which returns the potatoes to the ground for hand pickers.

The elevators run on a series of smooth rollers and agitators. The shaking action of the conveyors is varied by adding or removing agitators. Very little agitation is necessary when working in light dry soils. Some elevator diggers have rubber-covered rod links or webs to reduce tuber damage.

Working depth is controlled by a pair of depth wheels at the front of the digger. Careful setting of the share depth is important; potatoes will be cut if the share is too shallow and excessive soil will be put on the elevator if it is too deep. The depth wheels are usually combined with trash discs which clear a path for the share by cutting through weeds and tangled haulm.

Forward speed and power take-off speed are altered to suit crop and soil conditions. The object is to reduce tuber damage by retaining

soil with the potatoes for most of the way up the first elevator.

The power take-off drive is transmitted through a gearbox and roller chain or vee-belt to the elevators. A slip clutch protects the drive and the elevators from damage through overload or blockages.

Potato spinners

A flat share passes under the ridge to loosen soil and potatoes. The entire ridge is thrown sideways by either one or two spinning wheels against a net or canvas curtain at the side of the machine. The potatoes hit the net and fall to the ground, while most of the soil passes through the net.

Potato spinners are power driven, and mounted on the three point linkage. Correct working depth is important, because the potatoes will be buried if the share is too deep and cut if it is too shallow.

Using potato diggers

Maincrop potatoes normally have the haulm (tops) removed before lifting if they have not died off already. This is done with a mechanical haulm pulveriser—high speed power take-off driven flails under a metal canopy—or sprayed with a chemical defoliant. Early potatoes are often lifted with their tops intact.

It is common practice to leave the end headlands unplanted, to provide a space for turning with a tractor when spraying, ridging, etc. Some farmers plant the headlands and leave an unplanted strip further into the field for turning. This method gives a wider headland for manoeuvring after the outside rows have been lifted.

When harvesting potatoes with a digger, the headland is lifted first and then the main part of the field, usually in sections. After the first few rows have been lifted, a deflector can be fitted to the back of an elevator digger which guides the potatoes into a narrow band behind the machine. This provides space for the tractor wheels when lifting the next row, if the previously harvested crop has not been picked up.

Care must be taken to avoid damaging the potatoes by running the share too close to the surface. The share must not be set too deep either, for this will overload the machine and will bury many potatoes.

Elevator diggers can also be used, with some modifications, to lift other crops such as carrots, onions and flower bulbs. Conveyors with narrower link spacing and different shaped shares are needed for lifting some crops.

PLATE 18.12 *A single-row potato harvester in work. The row being lifted is to the right of the tractor, allowing it to run on land which has been cleared. Note the side elevator has been lowered to reduce damage to the crop.*

(Grimme)

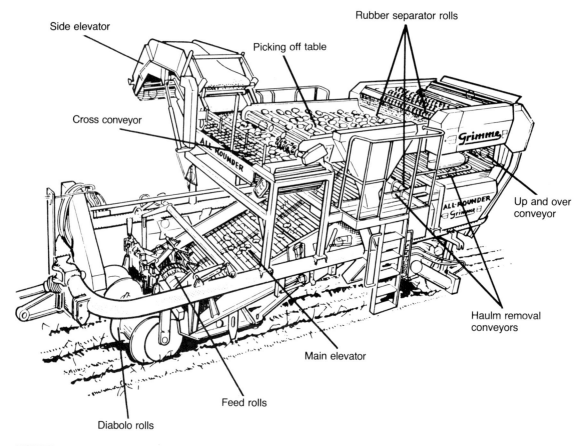

FIGURE 18.8   Potato harvester crop flow diagram.   (*Grimme*)

Maintenance of potato diggers

Frequent lubrication is important, but oil and grease should not be applied to parts which work in, or close to, the soil. Gearbox oil levels, tyre pressures and chain or belt tensions should be checked at regular intervals. Slip clutches should be checked for free movement at the start of the season.

Potato Harvesters

Potato harvesters lift either one or two rows at a time. They may be trailed, self-propelled or mounted around a tractor. Most harvesters have a platform for hand pickers who sort the crop from clods, stones and trash. After sorting, the potatoes may be elevated to a trailer, bagged up on the machine, or stored in a bulk hopper on the harvester which is emptied into a trailer when it is

full. Other crops such as onions and carrots can be harvested after fitting suitable lifting shares and elevator webs.

Manned potato harvester crop flow

Lifting methods are similar on most harvesters, while sorting mechanisms are rather more varied.

The diabolo roller runs over the ridge to steer the wide lifting share under the row of potatoes (see Figure 18.8). Feed rollers help to guide the crop onto the main rubber-covered web-type elevator where much of the soil is removed. Haulm extraction rollers grip the potato tops and return them to the ground. The potatoes, stones and clods are lifted to the sorting platform by an up-and-over elevator. After passing over a series of rubber separator rollers, which remove most of the stones and clods, the crop reaches the picking-off table.

PLATE 18.13   *An automatic bagger unit on a two-row harvester. It weighs out 25 kg, then cuts off the flow of potatoes while the sack is replaced. Some models have a sack stitcher.*   (Grimme)

There is enough room on the picking-off platform for up to six people to remove any remaining clods, stones or other rubbish. After hand sorting, the potatoes fall gently on to a cross-conveyor which carries them to the side elevator. The height of the elevator discharge point from the trailer floor is adjusted with a hydraulic ram to maintain a very short drop for the potatoes. In very good working conditions it is possible to use the harvester with very few or no hand pickers at all.

Hydraulics play an important part in the operation of the harvester. The machine is kept level and the pre-set working depth is maintained by hydraulic rams operated from the tractor cab. Rams are also used to raise and lower the lifting shares and control the discharge height of the side delivery elevator. A hydraulic steering kit, linked to the tractor, can be fitted to the harvester wheels. This improves manoeuvrability on headlands and keeps the machine on the row, especially in difficult conditions.

There is also a bagger version of this harvester. The side elevator is replaced with a bagging-off platform large enough for two or more people and a quantity of full sacks. Paper sacks are attached to the bagging-off spouts; when full the sacks are removed, weighed and tied. Some harvesters have an automatic weigher, which cuts off the flow of potatoes to the sack when it reaches the required weight.

Unmanned harvester crop flow
This is a less complicated machine, without a picking-off platform. It harvests and mechanically sorts two rows at a time, then elevates the potatoes into a trailer (see Figure 18.9).

The diabolo rollers guide the wide lifting

PLATE 18.14   *Hydraulic steering arrangement on a two-row harvester.*   (Grimme)

share under the ridges. Potatoes, soil and tops move on to the main web-type elevator where most of the soil is shaken from the crop. A haulm extracting roller at the top of the elevator removes most of the vegetation. The crop moves on to a shorter conveyor and second haulm extractor. From here, the potatoes, large stones and clods pass over rubber separating rollers to a cross-conveyor. The potatoes are transferred to the side delivery elevator and conveyed into a trailer. The discharge height of the elevator is controlled with a hydraulic ram so that the distance the potatoes must fall is kept to a minimum.

This type of harvester can be supplied with a picking-off platform, but it is intended for unmanned operation in soils which are relatively free from stones and large clods. Ideal soil con-

ditions can be achieved by removing large stones and clods with a clod separator before the crop is planted (see page 204).

Rotary cage harvester crop flow
Another type of harvester employs a rotary cage to lift the crop from the main elevator up to the picking-off table (see Figure 18.10).

The share lifts the ridge and haulm on to the main elevator. When the crop reaches the top of this elevator, with much of the soil shaken out, it falls through the haulm extracting conveyor to the rotary drum conveyor.

The haulm extractor returns any green material to the ground. The drum conveyor is a cage, made from heavy steel mesh and covered with rubber to reduce tuber damage. It has divisions inside it which form compartments for carrying the crop

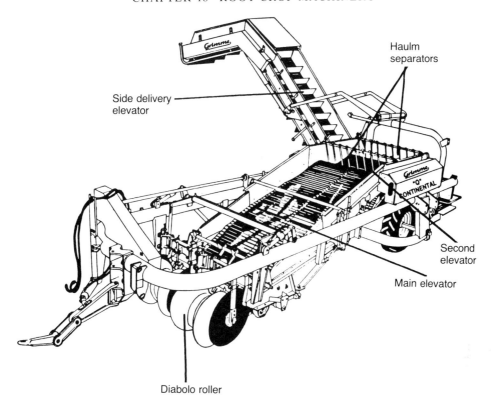

FIGURE 18.9 Unmanned two-row harvester. (*Grimme*)

PLATE 18.15 *Two-row harvester with diabolo rolls for depth control. Note the trash discs and the closely spaced bars on the web elevator.* (Grimme)

up to the short belt conveyor. Potatoes, clods and stones fall from this conveyor on to an inclined separator belt with rubber spikes.

Soil and small stones are trapped between the spikes and are returned to the ground. The potatoes run down the spikes to the picking-off table. Hand pickers remove any remaining stones, clods or damaged potatoes from the picking-off table and put them on a narrower conveyor running alongside, which returns the rubbish to the ground. The potatoes pass from the picking-off table over a rubber separating spool which removes any remaining soil. The process is completed by another conveyor which carries the potatoes to a side delivery elevator or bagging-off platform.

Maintenance and use of potato harvesters
Maintenance requirements vary from one machine to another. Important tasks include

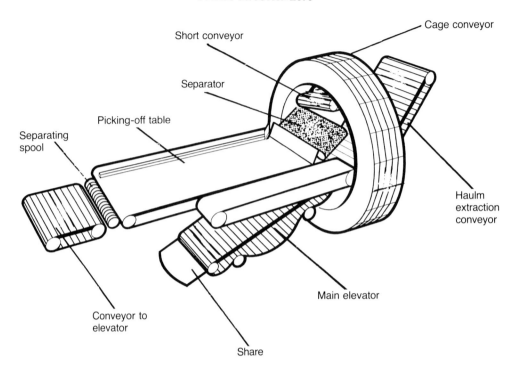

FIGURE 18.10   Manned harvester with rotary cage conveyor.

lubrication, checking gearbox oil levels, tyre pressures and correct tension of belt and chain drives. Overload clutches should be checked for operation and correct tension must be maintained on crop conveying elevators.

Harvesters are used in much the same way as elevator diggers. Wide headlands are needed for turning with these long machines.

Most growers remove the haulm before harvesting, either by spraying the tops to kill them or by chopping them up with a haulm pulveriser. Some maincrop potatoes will not be harvested until the haulm has died off.

Haulm Pulverisers

These power take-off driven machines can be front mounted, attached to the three-point linkage or trailed. A haulm pulveriser has a high speed flail rotor across the full width of the machine which will deal with two or four rows of potatoes at one time. Different lengths of flail are used to give a cutting profile which pulverises the haulm on the ridges and in the furrows without disturbing the potatoes. Some models have steel blades, others have rubber flails. The arrangement of the various length flails or blades can be altered to suit different potato row widths.

Wheels or skids can be adjusted to give a constant working height, which ensures minimum damage to the ridge when pulverising potato haulm. A typical two-row pulveriser requires about 45 kW (60 hp); the rotor speed can be varied from 1,000–1,850 rpm. Speed is altered by changing the size of pulley used in the vee-belt drive from the gearbox to the rotor. A slip clutch protects the drive from overload.

Other crops can be treated with a haulm pulveriser. Carrot tops, for example, can be removed after fitting a suitable blade formation to the rotor shaft. Some models of haulm pulveriser can be converted for chopping windrows of straw left by a combine harvester.

PLATE 18.16 *Front-mounted haulm pulveriser at work. This leaves the rear linkage free to pull a harvester.*

(Standen)

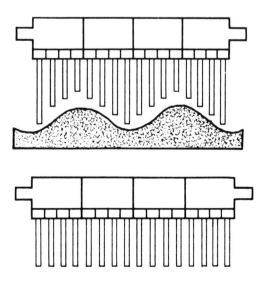

FIGURE 18.11 Haulm pulveriser blade arrangement for potatoes (*top*) and carrots (*bottom*).

Suggested Student Activities

1. Look for different types of sugar beet harvester. What topping, lifting and cleaning mechanisms are used?
2. Find out how a sugar beet harvester topper is adjusted.
3. Inspect the tops and roots after harvesting to see if you can find evidence of under-topping or over-topping.
4. Study an automatic potato planter to find out how the feed mechanism works.
5. Look at a complete potato harvester and trace the path of the potatoes from the ridge to the trailer or bag. Pay particular attention to the way in which the design of the machine keeps tuber damage to a minimum.

Safety Check

Avoid wearing clothing such as long flapping overcoats, loose belts and long scarves when working with farm machinery. Any item of clothing caught by a shaft, elevator link or similar mechanism will have serious, if not fatal, results. Footwear should have safety toecaps and non-slip soles.

Chapter 19

FARMYARD AND ESTATE MACHINERY

All farms will have machinery for the storage of grain, the preparation of animal food or the maintenance of hedgerows, ditches and fences.

Electricity is used to operate much of the grain storage and food preparation equipment. This makes it suitable for automatic operation, especially at night when cheap rate electricity is available.

ROLLER MILLS

These are used to crush or flatten grain. Rolled barley is used to fatten large numbers of beef cattle. A roller mill consists of two smooth steel rolls held slightly apart by spring pressure. A screw adjuster is used to move the rolls closer together against the springs. A hopper above the

PLATE 19.1   *A telescopic loader with a large capacity bucket loading grain from a floor storage building.*   (JCB)

rolls feeds grain between them. Rate of feed is controlled by an adjustable slide at the bottom of the hopper. The degree of crushing depends on the clearance between the rolls. A sieve in the base of the hopper collects any stones or other solid objects in the grain to prevent damage to the rolls. Once crushed, the grain is discharged under the mill for further handling or it may be conveyed direct to the cattle feed troughs.

The rolls will wear very quickly if they are allowed to run against each other without grain passing through them.

Very dry grain, below 17 per cent moisture content, is not in the best condition for rolling. The product will be rather dusty and there will be a tendency for the grain to shatter. Moist grain, stored in a sealed silo, will be in an ideal condition for crushing with a roller mill.

Roller mills are suitable for automatic operation. A cut-out switch in the hopper will turn off the power when the grain supply is exhausted. A flap switch is one form of automatic cut-out. The grain holds a metal plate on a pivot close against the inside of the hopper. When the hopper is empty, a counterbalance weight on the flap switch shaft raises the metal plate to a horizontal position. This movement operates a switch to disconnect the power supply to the mill.

A diaphragm or pressure switch is another way of achieving automatic control. A rubber diaphragm with electrical contacts inside it is pressed firmly against the side of the hopper by the weight of the grain. When the hopper is empty, the diaphragm moves away from the side of the hopper and disconnects the current.

Hammer Mills

Hammer mills vary considerably in size and output. The most common hammer mills have a power requirement of about 3.75 kW (5 hp). Much larger models are in use.

A typical hammer mill has a high speed shaft with a grinding rotor at one end and a fan at the other. It is usual to have eight flails attached to the rotor, their tips passing very close to a fixed circular screen. The screen has a large number of small round holes in it. Grain fed from a hopper to the flails (hammers) is beaten against the screen until the resulting meal is small enough

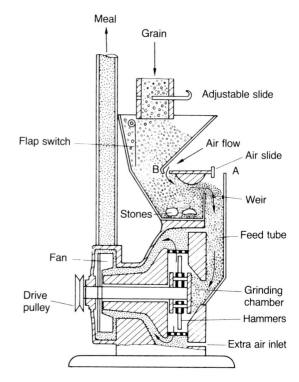

FIGURE 19.1 Hammer mill.

to pass through the holes. The fan serves two purposes. It draws grain from the hopper to the grinding chamber. After the meal has passed through the screen, the fan blows it to a bin or mixer.

The shaft carrying the hammers and the fan runs at speeds up to 6,000 rpm. It is driven by an electric motor with a vee-belt. An automatic cut-out switch in the hopper turns off the power supply when grinding is completed.

Adjustments
- *Degree of grinding* A hammer mill has a range of grinding screens. A typical mill will have screens with 1.5, 3, 6 and 9 mm holes. Other sizes are available. To change a screen, the grinding chamber cover is removed. This gives easy access to the screen.
- *Feed rate* The fan sucks air through the grain. The strength of the airflow through the grain determines how much grain is drawn into the grinding chamber. When the air slide is moved towards B (see Figure 19.1) most of the air is sucked through the feed tube without

passing through the grain. In this position there will be a low feed rate. By moving the slide towards A, less air can reach the grinding chamber through the feed tube. As a result more air is sucked through the hopper and an increased amount of grain is supplied to the hammers. There are extra air inlets in the grinding chamber cover. These are usually adjustable but are normally kept fully open. They supply extra air to the fan. This ensures all the meal is cleared away from the screen and blown to the bin. There is a risk of blockages in the grinding chamber if the air inlets are clogged with meal. It may be necessary to close the air vents slightly when grinding with a coarse screen to achieve full output.

Maintenance

Regular lubrication is important. High melting point grease is recommended for the high speed rotor shaft bearings.

The vee-belt drive must be at the correct tension to ensure that full power is transmitted to the mill.

The striking face of the hammers must be in good condition. On most mills, all four corners of the hammers can be used by reversing them and turning them from end to end. The rotor must be balanced to prevent vibration. If it is necessary to fit a new hammer, another must be fitted directly opposite to maintain rotor balance.

The screen can be reversed to obtain maximum life. When the four corners of the hammers are worn, the two faces of the screen will also be worn out.

Any stones or other solid objects in the grain will be trapped in the bottom of the hopper below the weir. These should be removed after a batch of grain has been milled.

Operation

A hammer mill must be allowed to gain full working speed before feeding any grain to the hammers. The automatic hammer mill is usually installed under a hopper and will grind a batch of grain without supervision. This allows the use of cheap rate electricity and the automatic switch in the hopper will turn off the power when the job is complete.

The air used to convey the meal to the bin or mixer must be removed from the product. There are two methods. A cyclone can be placed above the bin or mixer. This is a conical-shaped container with its point at the base and is at least one metre high. When the meal and air reach the cyclone, they whirl round inside the cone. The heavier meal is thrown outwards and falls through an outlet at the bottom of the cyclone. The air leaves from an opening at the top of the cyclone and is piped out of the building.

Air can also be separated from the meal with filter bags on top of storage bins or connected to food mixers. The filter bags allow the air to escape and the meal is retained inside the filter.

Tractor-driven Hammer Mills

High output hammer mills driven by a tractor belt pulley are used by some farmers, making it possible to grind grain in outlying stock units where there is an inadequate supply of electricity.

The power required can vary from about 15 kW (20 hp) for a small tractor-driven model, up to 38 kW (50 hp) for the larger mills. After grinding, the meal is separated from the air by a large cyclone.

Food Mixers

There are two types of mixer, vertical and horizontal. The latter consists of a metal trough with rotating paddles driven by an electric motor. There are not many horizontal mixers in use; the ability to handle wet material is their main advantage.

It is common practice to install a combined mill and mixer. This consists of a grain hopper, hammer mill, filter bags to remove the air and vertical mixer. In addition to the freshly ground meal, other ingredients are added to make up the required ration.

Vertical Mixers

This type of mixer has a large diameter vertical auger, running inside a tube in the centre of a cylindrical hopper which has a conical base. The auger, driven by an electric motor, runs at a speed of 250–400 rpm, depending on model.

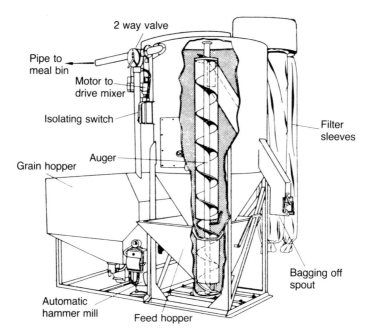

2 way valve

Pipe to
meal bin

Motor to
drive mixer

Isolating switch

Grain hopper

Auger

Filter
sleeves

Bagging off
spout

Automatic
hammer mill

Feed hopper

FIGURE 19.2   Mill mix unit.

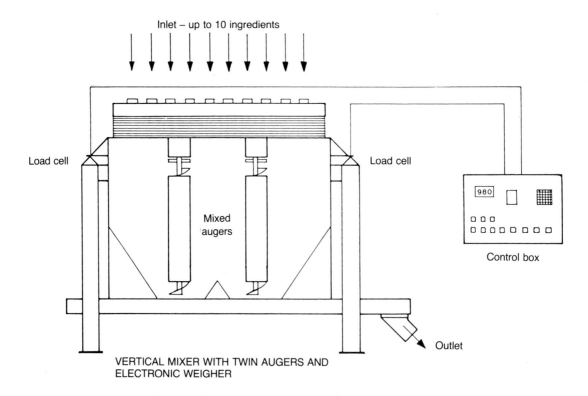

Inlet – up to 10 ingredients

Load cell

Load cell

980

Control box

Mixed
augers

Outlet

VERTICAL MIXER WITH TWIN AUGERS AND
ELECTRONIC WEIGHER

FIGURE 19.3   How a weigh mix unit works.   (*Ben Burgess*)

PLATE 19.2 *Food mixer with load cells which weigh each ingredient as it is supplied to the hopper. The conveyor at the base carries the mixed batch to a hopper or feed trough.* (Ben Burgess)

It mixes the ingredients either blown in by a hammer mill or tipped in from bags.

When the mixer starts up, the auger lifts the ingredients to the top of the tube and then allows them to fall back to the bottom of the hopper. This process continues until mixing is complete. The contents are then emptied from the unloading spout into bags or are conveyed direct to feed troughs, depending on the system of feeding used.

A typical vertical mixer holds 1 tonne and will mix a complete load in 20 minutes. Models from 500 kg to 2 tonnes capacity are available.

Accurate weighing of the separate ingredients is an important part of mixing rations and some mixers have an automatic weigher. The hopper is suspended on load cells which electronically weigh each ingredient as it is added to the mix. The machine shown in Figure 19.3 has the capacity automatically to add and weigh up to ten separate ingredients and then complete the mixing cycle. The control system's memory can store the recipes for ten different rations. When the mix is completed, it is discharged from the hopper discharge spout to a conveyor or feed wagon.

Grain Drying and Storage

Freshly harvested grain may have a moisture content of anywhere between 12 and 25 per cent. During most harvests, grain will come off the combine at 14–18 per cent moisture content. For safe, long period bulk storage, the moisture content for wheat, barley and oats should be no more than 14 per cent.

Many farmers have a combine harvester capable of cutting a far greater area of grain than is actually grown. This makes it possible to delay most of the harvesting until the grain is dry enough to store direct from the combine. In a wet season, some drying will be necessary but a high capacity harvester will reduce the amount of expensive grain drying needed.

Grain with a high moisture content will heat and, before long, moulds will appear. In this condition the grain is quite useless. The organisms which cause grain to heat up are always present but they can only work and multiply under certain circumstances. They must have:

- Oxygen.
- A reasonably high temperature.
- Sufficient moisture.

In order to store grain safely for long periods, the farmer must reduce the level of one of these to a point below which the destructive organisms cannot work.

There are three ways in which grain can be prepared for prolonged storage:

1. *Storage of moist grain in an airtight silo* to exclude oxygen. This uses the same principle as grass stored in an airtight silo to make silage. Moist grain from the combine will still be in that state when it is removed from storage, making it ideal for crushing with a roller mill.
2. *Chilling the grain* to bring it down to a very low temperature. A fan is used to blow air, cooled by a refrigeration unit, through the stored grain.
3. *Drying the grain* to reduce the moisture content. This is done by blowing warmed air through the grain, either in a continuous drier or after storing it on a ventilated floor or in bins.

Grain Driers

There are many designs of grain drier. Some use reasonably high temperature air to remove the moisture quickly. Others use air heated to a few degrees above ambient temperature (the natural air temperature), the drying process taking place over a period of days.

Grain can be conditioned by blowing air, at ambient temperature, through it when weather conditions are *suitable*. Blowing air through a bin full of grain on a damp or misty day will only increase the moisture content of that grain.

Damp air has a high relative humidity, so has little or no drying effect at all. However, air with a low relative humidity has the capacity to remove moisture from grain. Heating the air used for drying will reduce its relative humidity and increase its capacity to remove moisture. This is the working principle of a high temperature drier.

It is important, when using a grain drier, to know the relative humidity (RH) of the air blown through the grain. This can be measured with a hygrometer; it has a dial or digital read-out which registers the percentage RH of the air. When grain is dried, the relative humidity of the drying air increases. Tables are available which show the lowest grain moisture content it is possible to achieve at any given level of air temperature and relative humidity.

The moisture content of a grain sample can be found by using a moisture tester. Some testers work on whole grains, others require the sample to be ground by a small machine resembling a domestic coffee grinder. The moisture content is calculated by measuring the resistance of the sample to the flow of an electric current. The instrument converts the resistance reading to a moisture content percentage on a dial gauge or digital read-out.

High temperature driers

The most common type of drier consists of a vertical tower, two fans and a heater, usually oil fired. The damp grain is elevated to the top of the drying tower after it has been cleaned. The grain is allowed to fall very slowly. The sides of the drier are perforated and the heated air is blown through the grain. The bottom section of the drying tower receives air from the second fan. This blows unheated air through the grain to cool it. Once dried and cooled, the grain leaves the drier and is conveyed to bins or stored in bulk on the floor of the grain store.

The maximum temperature of the air used for drying is important. Grain for seed or malting must not be heated above 49°C, provided that it is under 24 per cent moisture content when harvested. Temperatures above this level will impair germination. Milling wheat can have an air temperature of 65°C and feed corn up to 82°C.

Another type of high temperature drier has a continuous horizontal conveyor. First hot and then cold air is blown through the grain on the conveyor. This type requires a much larger floor area.

Low temperature driers

There are two main low temperature drying systems—ventilated bins and ventilated floor storage. Before loading low temperature drying floors or bins, the grain is usually put through a pre-cleaner which, with the aid of sieves and an air blast, removes weed seed, chaff and other unwanted matter. If pockets of green material, chaff, etc. are allowed to collect, they will not dry properly and will cause hot spots in the stored grain.

The ventilated bin system of drying has slightly warmed air blown upwards through a perforated floor. Drying takes place gradually from the floor upwards, with moisture content being reduced by about 1 per cent per day. The bins are usually filled from the top with an auger and emptied with another auger from a small pit in the floor.

Ventilated floor storage may either have a perforated floor or air ducts placed at intervals across the floor of the building. The air chambers under the floor or the above-floor ducting are connected to a central duct where it is possible to control which section of the floor has warm air blown through it at any time. It is important to keep the stored grain at an even depth over the floor to ensure even drying. Overhead elevators with remote control systems enable the operator to fill any section of the floor. The store is usually emptied with either a sweep auger or a grain bucket on a tractor or fork lift truck.

A sweep auger has a horizontal auger in an open fronted trough which is gradually moved forward into the heap of grain. The collecting auger, with left- and right-handed flights, feeds the grain to a central high capacity elevating auger which loads it into a lorry or trailer.

High speed loading of floor stored grain can be achieved by a tractor loader fitted with a grain bucket. Telescopic loaders have the added

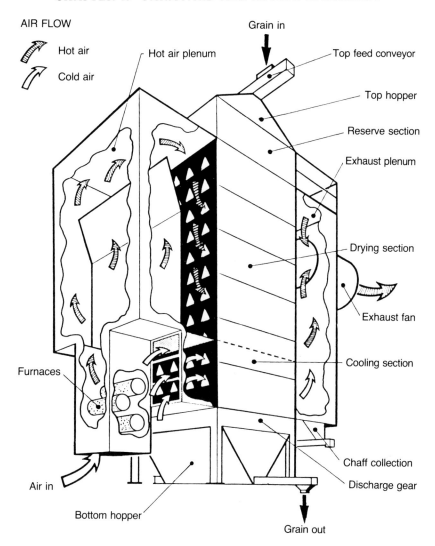

FIGURE 19.4 Section through a high temperature grain drier. (*Carier*)

advantage of both high and long reach when filling high sided lorries very quickly.

Portable driers
Fully portable grain driers, which are towed by a tractor, can be taken to remote storage buildings or used to back up existing drying facilities. Some models are completely self-contained and require no electricity. A combination of tractor power take-off shaft and gas or diesel to operate the drier furnace provide all the necessary services.

There are two types of portable drier: one dries batches of grain, the other is a continuous flow machine.

The batch drier is filled and the grain is continuously circulated by a central auger which lifts it from the bottom to the top of the hopper in a similar way to a vertical mixer. Warm air is blown by a power take-off or electric motor-driven fan to a warm air chamber at the centre of the drier. From here, the warm air passes through the perforated walls of the drier hopper and the grain, gradually reducing the moisture content. When

PLATE 19.3   *A large capacity grain bin storage system.*

the grain is dry, the auger empties the hopper into a trailer or conveyor system.

There are numerous models of portable batch drier. A typical medium size machine has a hopper capacity of 9 tonnes and requires a 30 kW (40 hp) tractor to drive the augers and fan. When reducing the moisture content from 20 to 15 per cent it will dry approximately 4.5 tonnes per hour.

Continuous flow driers are top filled and a thin wall of grain makes its way slowly downwards between two walls of perforated metal. Warm air is blown through to dry the grain, which is then cooled by a second fan which blows ambient air through it. The dry grain is discharged at the bottom of the machine. An example of a continuous portable drier has a holding capacity of 3.5 tonnes and can dry up to 6 tonnes per hour. The air is heated by a gas burner.

Grain Conveyors

Conveyors are used to move grain either vertically or horizontally. Many grain stores have remote control systems to start and stop the conveyors and micro-switches to protect them from damage through overloading or blockages.

Grain conveyors should have sufficient capacity to keep the intake pit empty enough to take the grain brought to the store from the combine. They must also keep the drier supplied with enough grain to keep it working at full capacity. Average combine harvester output is about 12–15 tonnes per hour and drier capacity is around 20 tonnes

PLATE 19.4 *A portable recirculating batch drier. The 450 mm wall of grain is contained between the inner warm air chamber and the outer perforated wall. The circulation and unloading auger can be seen above the centre of the drier.* (Opico)

per hour. Actual outputs will, of course, depend on the size of the equipment.

Belt and bucket elevators are used for vertical grain conveying. An endless belt with closely spaced buckets picks up grain from a hopper at the base and carries it to a discharge spout at the top. Their main use is carrying grain from the intake pit up to a pre-cleaner and drier. Bucket and elevator conveyors in farm grain stores have a capacity of 20–60 tonnes per hour, depending on the size and speed of the cups.

Chain and flight conveyors are used for horizontal conveying. An endless chain with closely spaced flights moves the grain along on the floor of the conveyor trough. Discharge openings at various positions along its length make a chain and flight conveyor suitable for filling a floor store. Most installations have remote control shutters at the discharge points. Chain and flight conveyors used in farm grain stores have capacities ranging from 20–60 tonnes per hour, although models with much higher outputs are made.

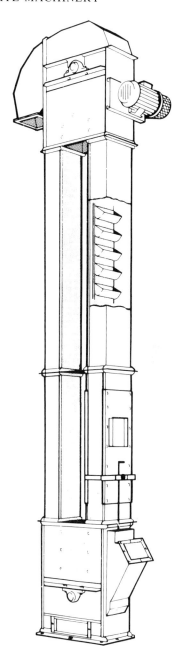

FIGURE 19.5 Belt and bucket elevator. *(Carier)*

Belt conveyors are used to move grain horizontally. They are smooth running and ideally suited to high level installation, especially when conveying over some distance. Grain is moved at high speed on an endless rubber belt driven and supported by rollers. The grain is fed to

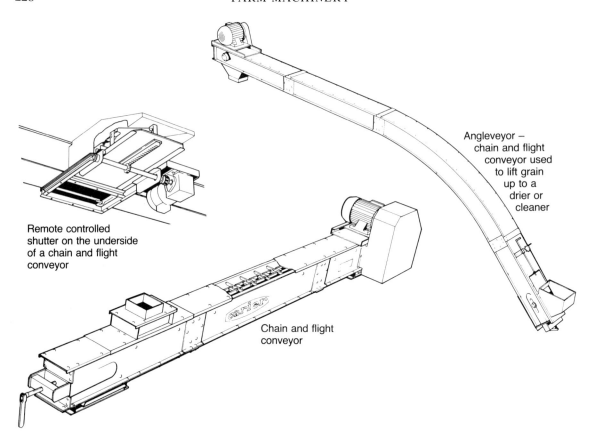

Angleveyor –
chain and flight
conveyor used
to lift grain
up to a
drier or
cleaner

Remote controlled
shutter on the underside
of a chain and flight
conveyor

Chain and flight
conveyor

FIGURE 19.6   Chain and flight conveyors.   (*Carier*)

the conveyor belt by a funnel-type chute and removed by a rather complicated discharge spout. This discharge mechanism makes belt conveyors rather less convenient than the chain and flight type for filling floor stores. The capacity of belt conveyors, when handling grain, is similar to that of chain and flight conveyors.

Augers have the advantage of being portable. The smaller models can be placed in position by hand, while the larger ones have their own chassis and wheels. Augers are mainly used for moving grain over short distances and emptying storage bins. Sweep augers are often used for emptying floor stores. Augers vary from 100–400 mm in diameter, the larger diameter models being made in lengths of up to 12 m. The capacity of a grain auger depends on its diameter and working angle. A 150 mm diameter auger working horizontally, for example, will handle

50 tonnes per hour, but when operating vertically, the output falls to 20 tonnes per hour.

HEDGE AND DITCH MAINTENANCE

Hedge Cutters

Most hedge cutters consist of a high speed flail rotor at the end of a hydraulically controlled arm which can be set at various angles to cut the sides and the tops of hedges. The rotor can also be positioned to cut the sides of ditches and banks. Another type of hedge cutter has a reciprocating knife cutter bar, which can be set at various angles with hydraulic rams. This is used by some farmers for hedge trimming; it will produce a very neat finish but cannot handle thick material.

The rotor on a flail hedge cutter is driven by

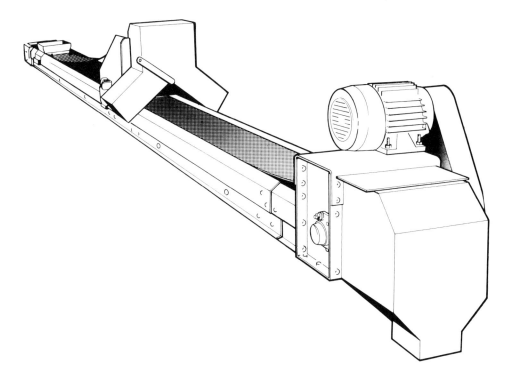

FIGURE 19.7   Belt conveyor.   (*Carier*)

a hydraulic motor, and the rams used to set the angle of the cutting head are controlled from the cab. The rotor has either one or two speeds, depending on model. A typical flail hedge cutter has a rotor speed of 3,000 rpm for hedge cutting and 2,400 rpm for grass.

Some hedge cutters have their own self-contained hydraulic system. Others use the tractor hydraulics for the rams and have a separate circuit for the hydraulic motor. Hedge cutters with an independent hydraulic system have an oil reservoir on the machine. The tractor power take-off drives the pump which supplies both motor and rams. The large tank at the right-hand side of the hedge cutter shown in Plate 19.5 is the oil reservoir for its hydraulic system.

Most hedge cutters have a rotor with a cutting width of about 1 m, while others have a 1.5 m rotor. The reach—the maximum distance from the tractor to the rotor flails—will depend on the length of the flail rotor arm. A general purpose model, with a two-part arm, has a reach of about 5 m. Hedge cutters with three-part arms can have a sideways reach in excess of 7 m.

PLATE 19.5   *A flail hedge cutter.*   (Bomford)

A break-away device is built into the flail rotor arm to protect it from damage when it hits an obstruction. Some models have a break-away which can operate in both forward and reverse directions.

Ditchers

Many large farms have an industrial tractor digger with a heavy duty front loader bucket and trench digger at the back. There are also numerous models of tractor-mounted ditch digger. The smaller machines are fitted to the three point linkage and use the auxiliary hydraulic connections on the tractor to supply oil for the rams. Others have an independent hydraulic system with a power take-off driven pump and an oil reservoir attached to the ditcher frame.

When in work, the machine is supported by hydraulically lowered legs. A set of remote control valves are used to move the ditcher arm up and down, also from side to side. The ditcher arm can swing through a working arc well in excess of 180 degrees. The position of the two-part arm above and below ground level is controlled by two rams, with a third operating the digging bucket. There is a wide selection of buckets, ranging from a very narrow one to dig trenches for water pipes up to a ditch cleaning bucket with a width of 1.5 m or more.

A typical farm ditcher can excavate a trench up to 2.7 m deep and has a maximum reach of 4.25 m from the support legs to the tip of the bucket.

CHAIN SAWS

The chain saw is used on many farms, but is a very dangerous tool and should always be used with great care. Most farm chain saws are powered by a two-stroke single-cylinder petrol engine; many have a diaphragm carburettor. When cutting, the engine runs at speeds in the region of 7,000–10,000 rpm or more.

Various lengths of chain guide bar are available, ranging from 300 mm bars on electric saws up to 600 mm or more on large petrol-engined models. A typical chain saw for general farm use

PLATE 19.6 *A general purpose chain saw with a 43 cc (2.1 kW) two-stroke engine.*   (Sachs-Dolmar)

will have a guide bar from 380–530 mm long. The engine will be in the 2.1–3.6 kW (2.8–4.8 hp) range.

The most important factor in chain saw operation is keeping the chain sharp. It is working correctly when the blades throw out wood chips, a sure sign that the blade is sharp. A blunt chain makes very small wood chips or sawdust. Frequent sharpening is required, but only two or three strokes of the file are necessary on a well-maintained chain. The tension of the chain on the guide bar is also important. There is a risk of the chain coming off the guide bar if it is too slack, and if it is too tight the bar and drive sprockets will suffer from excessive wear.

Most saws have an automatic chain lubricating system. The oil reservoir must be kept topped up to ensure the chain is well lubricated. Some saws have a hand operated pump for oiling the cutting chain.

Safety

Various safety mechanisms, required by law, are built into chain saws. They include:

- *Chain brake* A device used to stop the chain; this minimises the risk of an accident due to kick-back (violent upward movement of the saw).
- *Rear handle guard* Protects the right hand if the chain breaks.
- *Chain catcher* A pin which catches the chain if it breaks, working in conjunction with the rear handle guard.
- *Anti-vibration system* This dampens the vibration of the saw to reduce fatigue and improve cutting accuracy.
- *Safety lock-out switch* This prevents accidental chain engagement.

Chain saw users are required by law to wear a safety helmet with ear defender, eye protection,

PLATE 19.7 *An ATV with a 15 kW single-cylinder four-stroke air-cooled engine. This is a two-wheel-drive version with carrying racks at the front and rear and a towbar for trailed equipment.* (Honda)

PLATE 19.8   *Silage trailer being filled by a forage harvester.*   (Claas)

protective gloves and leggings (or trousers made from a special material) and safety boots. All clothing should be close fitting.

Manufacturers of chain saws, and agricultural colleges, provide instruction in the care and use of chain saws.

FARM TRANSPORT

Many types of material need to be transported around the farm, and in this high speed age people too need to get from place to place as quickly as possible.

All Terrain Vehicles

All terrain vehicles, or ATVs as they are commonly known, are popular with many farmers as a means of getting round the farm because these machines can negotiate most ground conditions. As well as providing personal transport, ATVs are used to carry out light work on arable crops such as applying slug pellets to cereal crops when the soil is too wet for tractors. With their low pressure tyres, ATVs have a ground pressure of around 0.15 bar (2.5 psi). Some livestock farmers use ATVs to help with the management of stock, especially on outlying fields. Not only do they save a lot of walking, they can also be used with a small trailer to carry fodder, fence posts, etc.

ATVs have a single cylinder air-cooled engine. Power output varies from 7–18 kW (9.3–24 hp) or more, depending on the model. Both two- and four-wheel-drive versions are made. A 12 volt socket outlet provides power to operate equipment such as an electro-broadcaster used to apply slug pellets. Other equipment, such as trailed sprayers, fertiliser spreaders, power brushes and mowers can also be used with ATVs, driven by a separate, single-cylinder air-cooled engine fitted on the machine.

PLATE 19.9 *A monocoque grain trailer.* (AS Marston)

Trailers

Every farm has trailers and there are never enough of them. Mainly constructed of steel, although some are partly or completely made of wood, trailers have two or four wheels and usually a pick-up hitch ring for automatic attachment. Some trailers have a jaw-type hitch used with a swinging drawbar.

The law requires that trailers with a jaw-type hitch have a screw jack to raise and lower the drawbar. There are also regulations concerning the provision of braking systems when trailers are used on the highway. Full lighting must be fitted to trailers used on public roads after dark or in poor daylight. All new trailers must have direction indicators.

When towing wide or overhanging loads on the highway marker boards are required on the load to warn other road users. The regulations are complicated. You should study them to be sure you do not break the law when using a trailer or moving a wide load on the public highway.

The range of trailers is vast. There are tippers, three-way tippers, high lift tippers, flat beds, drop sides and monocoque—a one-piece steel body with an opening tailgate. Specialist trailers, and conversion kits for standard models, are made to transport silage, grain, bales, etc. Tailgates on high sided trailers used for grain, silage, sugar beet, etc. may be operated with hydraulic rams or manually. The load capacity of trailers ranges from 1 tonne up to 20 tonnes

PLATE 19.10   *A three-way tipping trailer.*   (AS Marston)

or more. The small models are made for compact tractors. The largest trailers, equipped with a full hydraulic braking system controlled from the cab, are used with high horse power, four-wheel-drive tractors. The general range of trailers, suitable for most of the transport work around the farm, will have a load capacity of 5–14 tonnes.

Electric Fencers

Mains and battery operated electric fencer units play an important part in the management of livestock on grass. Electric fences can be used to make permanent paddocks or divide pastures for temporary grazing.

Up to 19 km (12 miles) or more of single line fence can be electrified with either a mains or a 12 volt battery fencer unit. A lower output 12 volt battery model can energise a fence installation of about 5 km (3 miles).

The fencer unit transforms electricity supplied by the mains or a battery to a much higher voltage—between 5,000 and 10,000 volts, depending on model. It works in a similar way to an engine coil ignition system. A transistorised circuit breaker interrupts the flow of current in the primary low voltage circuit, resulting in pulses of high voltage current being induced in a coil in the secondary circuit. There are between 40 and 60 pulses of high voltage current produced

per minute. Some fencer units have a mechanical circuit breaker. The high voltage current (with a very low amperage) passes through the fence wire. Any animal touching it will earth out the current and receive a sharp shock.

The fence wire is carried on plastic or porcelain insulators fixed to wooden posts. Steel posts, about 10 mm diameter, with a plastic pigtail insulator, are often used to erect lengths of temporary fencing. For small livestock, such as sheep and young animals, several strands of wire will be needed or, alternatively, a special type of polythene mesh fence with wire woven into it can be used. A host of fencing accessories are available including gate handles, strainer posts and neon testers to check if the fence is working properly.

Tall vegetation must not be allowed to touch the wire because it will partially earth out the current and reduce the effectiveness of the fence.

Suggested Student Activities

1. Study the layout of a mill mix unit. Find out how the meal is separated from the air and how to change the hammer mill grinding screen.
2. Look round a farm grain store. What type of drying, heating, conveying and storage systems are used?

3. Find out how to operate a hedge trimmer or a ditcher. The instruction book should be your starting point.
4. Study the Farm Safety Regulations concerning the use of chain saws. List the protective clothing which must be worn when using a chain saw.

Safety Check

Never allow anyone to ride on the drawbar when towing a trailer. The only safe place for a passenger is on the trailer floor.

Chapter 20

FARM POWER

Farmers have various sources of power available to them. Power to work the land comes from the internal combustion engine. Electricity is another major source of farm power. Heat is provided by oil, gas and electricity for use in livestock buildings, for drying grain and glasshouse crop production.

Internal Combustion Engines

Both petrol and diesel engines are used to drive standby generators and fixed equipment in buildings.

Engine power is measured in horse power or kilowatts. Many tractor sales leaflets quote

PLATE 20.1 *A diesel engine provides 176 kW (240 hp) for this combine. A 132,000 volt electricity supply line can be seen in the background.* (Massey-Ferguson)

236

engine power in hp, though some manufacturers give a kW rating. One horse power is equivalent to 0.75 kilowatts. This means that an engine developing 100 hp can also be rated as 75 kW.

One horse power equals 33,000 ft lb per minute. This means that in theory, a 1 hp engine can move a load of 33,000 lb through a distance of one foot in one minute. Alternatively it can move a load of 330 lb through a distance of 100 ft in one minute. One kilowatt equals 1,000 metres per second (m/sec). There are approximately 10 newtons in 1 kg.

Tractor specifications often give three horse power ratings. These are:

Brake horse power

BHP is the power available at the engine flywheel. It is measured with a dynamometer with the engine on a test bed. This device applies a type of brake to the flywheel. Measurements are taken from a dial gauge and the results are converted into brake horse power.

Drawbar horse power

DBHP is the power at the drawbar. It will always be less than brake horse power because energy is used to drive the transmission system, hydraulics, etc.

Power take-off horse power

PTO HP will be less than brake horse power but greater than DBHP. It can be measured by coupling a dynamometer to the power take-off shaft. Large power take-off driven machines have a high power requirement and the power available at the power shaft has been increased considerably during recent years.

Some or all of the three horse power ratings listed above are given in tractor sales leaflets. A typical tractor might have the following horse power figures:

BHP	—	78 hp (58.5 kW)
PTO HP	—	72 hp (54 kW)
DBHP	—	60 hp (45 kW)

This means that a tractor described as 78 horse power has only 60 horse power at the

(*Note*: one metric horse power (i.e. ps and cv) = 75 kg metres per sec and from this it can be calculated that 1 hp = 1.014 metric hp. It is more usual to quote engine power in kilowatts when using the metric system.)

draw-bar to pull implements. This power will be reduced still further by wheelslip, tyre flexing and rolling resistance. When pushing a loaded wheelbarrow on soft ground, it is hard work because the wheel sinks into the soil: this is rolling resistance. The front wheels of a tractor have the same resistance to overcome when moving forward.

SAE, DIN or BS?

Tractor sales leaflets may quote brake horse power or kilowatts as an SAE, DIN or BS rating. These abbreviations indicate the testing standards used to calculate the power produced at the engine flywheel (BHP).

DIN is widely used in Europe. DIN (Deutsche Industrie Normen) 70020 is a test standard which assumes the engine is installed in a vehicle. Air cleaner, cooling fan, water pump and exhaust system are fitted. The test is carried out on an engine test bed and the power calculated is a net figure.

SAE is an American power rating. SAE (Society of Automotive Engineers) J270 121 is a test standard using a bare engine on a test bed. The equipment driven by the engine in the DIN test is not fitted. The SAE rating is a gross figure and will be higher than the DIN rating for the same engine.

BS is the British Standard. BS AU 141a:1971 is yet another test standard. This is basically intended for road vehicles, but with less strict requirements for smoke emission when applied to agricultural tractors. It is carried out under similar conditions to SAE, power rating is gross and is likely to be higher than the DIN figure for the same engine.

The use of internal combustion engines inside buildings presents a number of problems. Exhaust fumes must be piped away. The engine air cleaner must have its inlet point outside the building to avoid collecting vast amounts of dust. The fuel tank may need to be outside the building to avoid the risk of fire and all hot surfaces and moving parts must be fully guarded. The installation must comply with the Farm Safety Regulations which apply to stationary machinery.

Electricity

The problems of installing internal combustion engines in a building do not apply to electric motors. Indeed, the opposite applies because an electric motor must not be exposed to damp working conditions unless it is specially designed for the purpose.

Electricity is a major source of farm power. It is generated at a network of power stations which feed electricity into the main supply lines (the National Grid) at a pressure of 275,000 or 400,000 volts. It is reduced to 132,000 and then again to 33,000 volts by transformer sub-stations. A village may have a supply of 11,000 volts from a transformer on the 33,000 volt line. This is again reduced by pole transformers to 240 single-phase or 415 volt three-phase supply.

Mains electricity to farms is usually 240 volts single-phase supply; it is transmitted by two wires and is suitable for most farm needs. Some farms have three-phase supply. This is a combination of three separate single-phase supplies arranged to give a pressure of 415 volts. Three-phase supply has advantages where heavy electrical loads occur. It is very expensive to install three-phase power lines and many country areas do not have it. Farms with three-phase electricity may have grain handling, milling and food-mixing equipment working at 415 volts.

Mains electricity is AC or alternating current. It cannot be stored in batteries. Power station output is regulated to meet the demand of the consumers. Tractor generators supply direct current (DC) to the battery for use when required.

Electric power is measured in watts; 1,000 watts is a kilowatt. The rate of flow of current in a circuit is measured in amps. Circuits and electrical appliances have fuses to give protection from overload. The size of a fuse in amps depends on the amount of electricity flowing in the circuit. You can find the current flow in a circuit with this formula:

$$\text{Amps} = \frac{\text{Watts}}{\text{Volts}}$$

For example, an electric heater has a power rating of 1,000 watts and is connected to a 240 volt supply. The current flow is calculated in this way:

$$\text{Amps} = \frac{\text{Watts}}{\text{Volts}} = \frac{1,000}{240} = 4.16$$

When the number of amps and volts are known, the formula can be turned round in this way to find watts:

$$\text{Watts} = \text{amp} \times \text{volts}$$

Example: find how many watts can be carried by a 240 volt circuit with a 5-amp fuse:

$$\text{Watts} = \text{amps} \times \text{volts}$$
$$= 5 \times 240 = 1,200 \text{ watts}$$

In practice a 5-amp fuse would be used for a 1,000 watt appliance.

Most portable electric equipment is fitted with a fused plug. It is important to have the correct fuse for equipment. The formula given above can be used to find the correct fuse rating (amps) provided that the voltage and watts are known.

Fuse rating	Maximum loading (watts)	Typical use
2 amps	500	Inspection lamp Soldering iron
5 amps	1,200	Small bench grinder Tyre vulcaniser
10 amps	2,400	Heavy duty power tools Angle grinders Water heaters
13 amps	3,000	Small portable electric welders

Suggested fuse ratings in the table above apply to single-phase equipment. Three-phase power tools and equipment are wired to an isolator switch and have a fuse in the main supply fuseboard.

Replacing a fuse
When you need to replace or rewire a fuse:

- Always disconnect the power supply first.
- Replace the fuse with one of the same rating.
- Call in an electrician if the fuse blows again as soon as the power is turned on.

Residual Current Devices
Many electrical installations have Residual Current Devices (RCDs) instead of rewirable or cartridge fuses at the distribution board. When an overload or dangerous situation occurs, the RCD cuts off the supply in milliseconds, giving far

greater protection to both equipment and user. When an RCD cuts off the supply, disconnect the appliance and reset the RCD. If the RCD fails to reset or trips out again when the appliance is reconnected, it should be checked by a qualified electrician.

RCDs should be periodically checked; a test button is provided for this purpose. You should read and follow the instructions supplied with the RCD.

Care of electric motors

Totally enclosed motors are normally used for equipment which works in very dusty conditions. Electric motors in the farm workshop and food preparation areas do not always have full protection from dust. In such cases, give what protection you can to stop a build up of dust inside the motor. It can block the air circulation vent and in time the motor will overheat. Attention to the following points will help prolong the working life of electric motors:

- Wipe off any oil which may appear or be spilled on the motor. It will attack the insulation and attract dust.
- Keep the motor dry.
- Avoid overloading, as this will cause overheating.
- Lubricate according to the instruction book. As a general rule, very little lubrication is required.
- Make sure that the motor is secure on its mounting bracket. Vibration increases wear.
- Check that the pulleys are aligned. Misalignment will cause excessive wear on the bearings and the drive belt.

Wiring a Plug

The colour code

Always connect the three wires of an electric cable in this way:

- *Green and yellow* to the largest plug pin. This is the *earth* connection.
- *Brown* to the *live* terminal. This is the terminal connected to the fuse holder in a fused plug.
- *Blue* to the third terminal which is *neutral*.

(*Note*: you may find that older electrical equipment has a green cable for earth, a red cable for live and a black cable for neutral.)

To wire a three-pin plug:

1. Strip about 50 mm of the outer covering from the cable.
2. Secure unstripped cable in the cable grip.
3. Run the three coloured wires to their correct terminals. Cut each wire to leave about 10 mm beyond the terminals for making the connection.
4. Remove about 10 mm of covering from each of the coloured wires. Where each wire is made up with several strands twist them together for extra strength.
5. Bend the wires clockwise round the terminals so they will stay in position when tightening the terminal screws. Some terminals have a hole for the wire with a screw to secure the wire in position.
6. Fit the correct fuse, check your work and finally replace the plug top.

You will find some power tools have only two wires, one blue and the other brown. There is no earth terminal because the equipment is double insulated.

The cost of electricity

Electric power is measured in kilowatts. Consumers pay for electricity according to the number of units they use. One unit is used when a 1,000 watt appliance operates for one hour. A one kilowatt electric fire running for one hour will use one unit. You can calculate the cost of using electric equipment in this way:

Total cost =
kilowatts × time (hours) × cost per unit

Example: find the cost of running a 2 kW heater for 5 hours at 6.2 p per unit.

Cost = 2 kW × 5 hours × 6.2 p = 62 p

To find the cost of running an electric motor, the horse power must be converted to watts. In theory one hp is equivalent to 750 watts, but in practice it should be assumed that a one hp motor will use one unit per hour. This allows for higher power needs when starting and any losses in efficiency while running.

All electricity from the mains passes through a meter. This measures the number of units consumed. Domestic users pay for their electricity according to the number of units used and also pay a quarterly fixed charge. It is possible to have a second meter, to measure how much electricity is used during off-peak periods; there is a reduced charge per unit during that time.

Many farmers pay a block tariff. This means that the first block, or number of units used in each quarter, is charged at a high rate per unit and after that the price falls drastically. This is an advantage to the farmer who uses a great deal of electricity.

The safe use of electricity
Electricity can kill. The following points must be remembered when using or working near electrical equipment:

- Always use the correct fuse for the appliance.
- Protect light bulbs in farm buildings. Keep them clean and well away from any materials which could catch fire.
- Keep extension leads in good condition. Place them in a position where they will not be damaged. Never make do with any odd piece of cable as an extension lead.
- Never use unsafe electrical equipment. Report any faults to someone in authority as soon as possible.
- Do not attempt temporary repairs to electrical equipment or circuits. Call in a qualified electrician.
- Never work with electrical equipment when you have wet hands.
- Take great care when working near overhead cables with metal ladders or irrigation pipes. Make sure that there is plenty of room before taking high loads or machinery under low cables.

Water is a conductor of electricity. *Never* tackle a fire caused by an electrical fault with a water-based fire extinguisher or any other type of extinguisher which employs any form of liquid. Use a *dry powder* extinguisher. Turn off the power as soon as you can if you are ever the first person at the scene of a fire. If this is too risky, leave it for the fire brigade.

Electric shock
If you find someone who has received an electric shock, the first action must be to turn off the power supply. When this is not possible, free the person by wrapping insulating material round them such as dry cloth or rubber, or use the casualty's own clothing. Never touch the person's skin until the current has been turned off or the casualty has been dragged clear of the danger area. You could also get an electric shock if this advice is ignored. Once clear of danger apply artificial respiration if breathing has stopped.

Gas and Oil

Many farmers use electricity for heating. It is clean and trouble-free until there is a power cut. It is also one of the more expensive forms of heating.

Oil is an alternative fuel to provide heat, especially for grain-drying installations. Provided that the burners are serviced at regular intervals and fire precautions are observed, oil presents few problems as a heating fuel.

Gas is favoured by many farmers, especially those who keep livestock. It can be collected in cylinders from a local dealer or delivered by tanker to the farm where it is stored in permanent storage tanks.

Two gases, butane and propane, are in common use. They are both heavier than air and for this reason any leakage of gas will collect at low level. Permanent propane storage tanks must be sited outside a building. Portable installations such as heaters can have the storage cylinder inside the building or workshop. Take care when connecting gas cylinders to keep the threads clean. Any leak will smell, and if you suspect a leak turn off the supply at once. The source of a leak can usually be found by applying a small amount of soapy water to all joints: tell-tale bubbles will come from a leaking connection. When a leak is suspected, extinguish all flames and naked lights immediately.

The Transmission of Power

Power developed by internal combustion engines and electric motors can be transmitted to the driven machine in various ways. These include:

Vee-belts Commonly used to transmit drive from electric motors. The vee-belt drives with the sides of the belt which grip against the sides of the vee-belt pulley. The belt must not be allowed to run in the bottom of the pulley. Vee-belts are sometimes used with multiple vee pulleys; it is not unusual to have three or four belts running side by side where it is necessary to transmit a high horse power from the motor to the machine. Multiple drive vee-belts must be replaced in complete sets when they are worn.

Flat belts This type of belt is suitable for transmitting power over greater distances than a vee-belt. Pulleys for flat belts usually have a slight ridge, or crown, at the centre of the pulley face to help keep the belt in correct alignment.

Drive can be reversed by crossing the belt in the shape of a figure eight. Belt drives on implements and fixed equipment must be fully guarded both at the run-on points and along the entire length of the belt. There must also be an isolator switch or engine stop control at the working position to enable the operator to stop the drive to the machine.

Chain drives These are more suitable for slow speeds and give a positive drive with no possibility of slip. A slip clutch may be required to protect chain drives from damage through overload.

Roller chains are available in both high and low speed versions. Chains are made in many sizes and also as single, double row (duplex) or three-row chain (triple). Roller chains need frequent lubrication to ensure a long working life. The wheels which carry the chain are called sprockets. Sprockets and chains will wear at approximately the same rate. A worn chain has a lot of movement between the links and a worn sprocket can be recognised by its hook-shaped teeth.

The other type of chain in common use on the farm is the hook link type. It is intended for low speed work and the conveyor chain on the floor of a manure spreader is an example of its use. A hook link chain must be fitted with the hook leading. The chain is disconnected by slackening the chain tension and unhooking a convenient link.

Chain drives must be correctly tensioned. An idler or jockey sprocket is the most common method of tensioning a chain drive. This is a small sprocket on an adjustable arm which can be moved to take up any slack in the chain. When all adjustment has been used and the chain is still slack, a link must be removed from the chain.

Shafts
A shaft is a simple way to transmit drive over a distance. When the two ends of the shaft are not contained within a rigid framework, allowance must be made for any misalignment of the shaft. A tractor power take-off shaft has two universal joint bearings to allow the implement input shaft to be moved in relation to the tractor shaft.

Gears
Spur gears are only suitable to transfer power from one shaft to another running parallel to it. Gears can be used to increase or decrease shaft speed. A gear on a driving shaft which turns clockwise will turn a gear on a parallel shaft in an anti-clockwise direction.

Bevel gears are used to transmit drive through an angle from one shaft to another. The crown wheel and pinion in a tractor transmission system is an example of a right angle drive arrangement. A worm and wheel drive is another way of transmitting drive through a right angle.

Spur and bevel gears may have either straight or helical teeth. Helical teeth are at an angle and give a quieter and stronger drive than straight teeth gears.

Gear and pulley speeds
The speed of a driven gear can be calculated if the speed of the driving shaft and the number of teeth on both gears is known. Use this formula:

$$\frac{\text{Teeth of gear A}}{\text{Teeth of gear B}} = \frac{\text{rpm of gear B}}{\text{rpm of gear A}}$$

The same formula can be used for belt pulleys by substituting the number of gear teeth for the pulley diameter.

Example: find the speed of a 6 teeth gear driven by a 15 teeth gear which turns at 80 rpm. Using the formula, make gear A run at 80 rpm with 15 teeth, gear B has 6 teeth:

$$\frac{\text{Teeth on A}}{\text{Teeth on B}} = \frac{\text{rpm B}}{\text{rpm A}}$$

$$\frac{15 \text{ teeth}}{6 \text{ teeth}} = \frac{\text{rpm B}}{80 \text{ rpm}}$$

$$\frac{15 \times 80}{6} = \text{rpm of B}$$

$$\frac{1,200}{6} = 200 \text{ rpm}$$

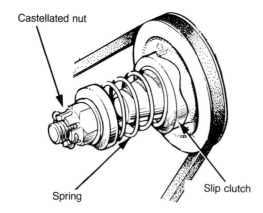

FIGURE 20.1   Serrated face slip clutch.

One type of slip clutch has two serrated metal faces, held together by spring pressure. Figure 20.1 shows this type of slip clutch in a vee-belt drive. When overloading occurs, the pulley continues to turn but the shaft does not. The movement of the serrated faces against each other makes a loud clattering noise.

Another type of slip clutch has friction-lined plates. The friction plates are held between smooth metal faces by spring pressure. In Figure 20.2 the power take-off shaft is firmly attached to the metal pressure plate. The friction plate is fixed to the driven shaft. When overloading occurs the power shaft still turns but the friction plate remains stationary. There is no noise from the clutch. Continuous slipping will overheat the friction plate.

An over-run clutch acts in the same way as a bicycle freewheel. It is spring loaded and has one square jaw and one angled jaw. The clutch will drive the machine when the power shaft is engaged. When the drive is disconnected the over-run clutch comes into action. It allows the drive shaft to stop while the machine slows down at its own pace. The over-run clutch makes a clicking sound as the jaws move against each other.

In this example, the speed of the driven gear is 200 rpm. The same method can be used to calculate the speed or diameter of a pulley provided that the other measurements are known.

Safety clutches
Chain drives and some belt drives are protected by safety clutches which will slip if the drive is overloaded.

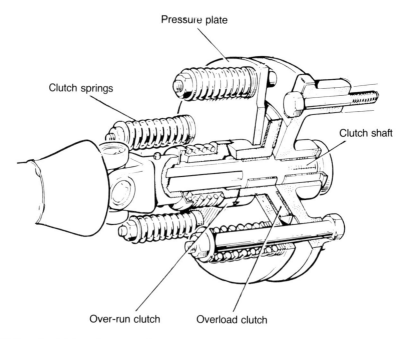

FIGURE 20.2   Friction slip clutch and over-run clutch.

Suggested Student Activities

1. Study sales leaflets to find the differences in BHP and PTO HP for various tractors.
2. Practise wiring a three pin plug.
3. Locate the mains electricity fuse box or RCD unit on the farm. Keep the area around them clear so they can be reached in an emergency.
4. Look for the different types of belt, gear and chain drive. Find out how the examples of chain and belt drive you find are tensioned.
5. Make sure you know the location of all the fire extinguishers on the farm. Read the labels to find out which types of fire each extinguisher may be used against.

Safety Check

Overhead electric cables in a farmyard or across fields must be treated with respect. Accidental contact with overhead lines, for example by the raised body of a high tipping lorry or trailer, by an elevator, or by irrigation pipes when moving them to a new position, can result in electrocution.

When ploughing or cultivating too close to an electricity post you may catch it, or the stay wire, with the implement, bringing down both post and wires. When this happens the cables, often still live, may finish up draped across the tractor.

Chapter 21

THE FARM WORKSHOP

MATERIALS

The more important materials used to manufacture farm machinery include various metals, both ferrous and non-ferrous, and certain types of plastic. Small amounts of timber, canvas and rubber are also used.

Ferrous Metals

Ferrous metals contain iron. There are many types of ferrous metal; of these, cast iron, carbon steels and alloy steels are used to make tractors and farm equipment.

The amount of carbon mixed with the iron

PLATE 21.1 *Pressure washing.*

during manufacture determines the type of iron or steel.

Iron ore, mined in many parts of the world including North America and Sweden, is put in a blast furnace with measured amounts of coke and limestone. The limestone acts as a flux which combines with impurities in the ore to form slag. The molten iron is drawn off from the blast furnace and run into sand moulds called pig beds. The resulting pig iron is the raw material for the production of various types of iron and steel.

Cast iron

Pig iron, coke and some limestone are put into a special furnace called a cupola. The pig iron is refined in the cupola and is then drawn off into ladles for transport to the casting shop. Here the molten iron is poured into sand moulds to produce the required shape. When cool the new casting is removed from the mould and after cleaning it is ready for the machine shop.

Cast iron is relatively cheap and is easy to machine and weld. It is a very brittle material but it does have considerable strength in compression (resists crushing). Cast iron is very fluid when molten and this makes it suitable for casting intricate shapes such as engine cylinder heads and cylinder blocks.

There is usually between 3 and 4 per cent carbon in cast iron. It also contains small amounts of silicon, sulphur, manganese and phosphorus. The actual quantities of these metals varies according to the type of cast iron. When the molten iron is allowed to cool naturally, grey cast iron is produced. Rapid cooling will produce white cast iron. This has a finer structure, is harder than grey cast iron and is rather more brittle.

Chilled cast iron is, in fact, white cast iron. It is used to make soil wearing parts such as ploughshares. A chilling plate is placed in the sand mould. This causes rapid cooling and gives a hard-wearing surface. A ploughshare has a hard underside and a soft top. This helps to keep the share sharp, because the top wears faster and keeps a good cutting edge on the share.

Uses of cast iron

The list of iron castings on farm machinery is endless: engine cylinder blocks, gearbox housings and rib-roll rings are a few examples.

Sometimes malleable iron castings are used when producing farm equipment. A malleable casting is one which has had special heat treatment to make it withstand shock. The heat treatment removes the brittleness. Some mower cutter bar fingers and plough fittings are malleable castings.

Wrought iron is another ferrous metal. It is almost pure, with only about 0.1 per cent carbon. It is very expensive and no longer used for agricultural purposes. You may still find the odd chain, shackle or wood splitting wedge made of wrought iron in a corner of the farm workshop. This metal gives warning of breakage. Instead of suddenly snapping, wrought iron will stretch and the skin peels away to reveal a slate-like structure.

Steel

The basic difference between iron and steel is the carbon content. The amount of carbon in steel will vary from 0.1–1.5 per cent. Steel becomes harder and more brittle as the carbon percentage increases.

Low carbon steel (mild steel)

This is the cheapest and most common form of ferrous metal used to make farm machinery. Mild steel contains from 0.1–0.3 per cent carbon. It can be machined, welded, bent and sawn with ease. It has no great strength and will rust very quickly if it is not protected from the weather. Painting, galvanising (coating with zinc), tin plating and plastic coating are some of the methods used to prevent corrosion of mild steel.

Uses Guards, tractor bonnets, machine frames and farm gates are a few examples of the wide use of low carbon steel. It can be purchased from metal merchants in many shapes including sheets, tubes, square, round and flat sections.

Medium carbon steel

Although more expensive, medium carbon steel is harder and tougher than mild steel. It contains between 0.3 and 0.7 per cent carbon. It can be welded and forged. Medium carbon steel can also be machined, drilled and sawn but its increased hardness makes these processes more difficult. It can be annealed or softened by heat treatment to make machining easier.

Uses include such items as plough beams, axles, plough discs and cultivator tines.

High carbon steel
The carbon content ranges from 0.7–1.4 per cent. High carbon steel tends to be rather brittle. It is a hard material and is used to make, among other things, chisels, punches, files and twist drills.

Alloy steel
An alloy is a mixture of different metals. Alloy steel has small amounts of one or more non-ferrous metals added to the steel and carbon. The added metals give the steel special qualities. Examples of alloy steels include:

- *Stainless steel*, which is made by adding chromium, is used where corrosion must be avoided. Dairy equipment is one example of the use of this form of alloy steel.
- *Tool steel* has tungsten added in very small quantities to give a good cutting edge.
- *Nickel steel* is very tough and has the necessary strength for making axles.
- A mixture of tungsten and manganese added to the steel gives increased wear resistance.

Heat Treatment of Steel

Low carbon steel is a relatively weak material and because of its low carbon content it cannot be fully hardened. The outer surface can be hardened by a process known as case hardening. Steels with a higher carbon content can be hardened and tempered or annealed to soften them.

Hardening
Steel can be hardened by heating it to red heat and then cooling it rapidly by quenching it in water. Some factory hardening processes use oil as a coolant. The action of quenching gives a hard metal but it will be rather brittle.

Tempering
This process is used to remove the brittleness from hardened steel. The metal is reheated to a certain temperature and quenched. The colour of steel changes as its temperature increases—from the natural colour of polished steel to straw, red, brown, purple and blue. The colour of the metal is used to decide when to quench.

A chisel is tempered by polishing the cutting edge with emery cloth and then applying heat on the handle above the cutting edge. The temper colours will slowly appear. A chisel should be quenched when the colour above the cutting edge turns to dark purple.

Annealing
This process is used to soften metals to make them easier to work or machine. The steel is heated to red heat and allowed to cool very slowly.

Case hardening
Mild steel cannot be case hardened in the same way as steel with a higher carbon content. However, it can be case hardened by adding carbon to the surface of the metal to give it a hard outer skin or case. To case harden mild steel, it is heated to a bright red colour and then dipped into a carbon rich powder. The metal is then reheated to a bright red colour followed by quenching in clean cold water. The quality of hardening can be tested with a file. The process should be repeated if the surface is not fully hardened. Case-hardening powder can be purchased from most tool shops.

Non-ferrous Metals

There are many non-ferrous metals but very few of them are used in the manufacture of farm machinery. A number of non-ferrous alloys are also used to make farm equipment. The more important non-ferrous metals and alloys include:

Copper
This metal is a good conductor of heat and electricity. It can be welded, is easy to work and resists corrosion. Copper is a soft metal but has a strange property of work hardening. This means that copper will become hard with age or if it is consistently hammered or bent. It can be softened by annealing. The copper is heated to a cherry red colour and then quenched in clean cold water.

The more important uses of copper include water pipes and fittings, electrical components and some engine gaskets.

Zinc
This corrosion-resistant metal is used to coat steel sheets, tanks and pipes. The process of

coating steel with a thin covering of zinc is called galvanising.

Lead

Lead is used for making the plates in tractor batteries. It is a very heavy metal. Freshly manufactured lead has a bright grey colour. It soon forms a surface coating of lead oxide which gives the dull grey colour. This coating protects the lead from further corrosion.

Solder is an alloy of tin and lead. It is used to repair electrical equipment, plumbing installations, etc.

Aluminium

This is a very light material of moderate strength. In its pure form aluminium is used mainly as sheeting. When aluminium is alloyed with other metals, including copper, zinc and silicon, it is used to make castings. Examples include pistons, small engine cylinder blocks and gear casings.

Brass

Brass is an alloy of copper and zinc. There are two main types of brass. Soft brass contains about 40 per cent zinc, it is easy to work with and is produced as sheets, tubes, wire and other shapes. Hard brass contains more than 40 per cent zinc and is used to make brass castings. Examples include sprayer parts, taps and electrical components. It has a high resistance to corrosion. Brass is a good material for making bearings.

Bronze

There are several types of bronze. It is basically an alloy of copper and tin with small amounts of other metals added to give special properties. Phosphorus is added to make phosphor-bronze, which is a good bearing material.

Identification of Metals

The basic non-ferrous metals are easily recognised by their colour but iron and steel require more knowledge to identify them.

The hardness of the metal can be determined with a hacksaw or file. Low carbon steel (mild steel), grey cast iron and the non-ferrous metals are easy to cut. Medium or high carbon steels are much more difficult. The grained structure of cast iron will soon distinguish it from mild

steel. Another test to identify the types of iron and steel requires an electric grinder. The sparks from the different metals are a good aid to identification.

- Low carbon steel produces large quantities of white, spear-shaped sparks and a few stars.
- Medium carbon steel has rather less sparks. They are white with a lot of star-shaped sparks mixed with the longer spears.
- Cast iron gives a small quantity of reddish sparks which turn to a straw colour.

A hammer and a hacksaw can also be used to help with identification. Saw about one-third of the way through a sample of metal about 12 mm thick. Place it in a vice and hit the metal near the cut with a 1 kg hammer.

- Low carbon steel will bend over some distance before it breaks.
- Medium carbon steel will bend a little before breaking.
- Cast iron will snap off with very little effort.

Plastic

Plastic does not corrode. This property makes it ideal for many agricultural applications, especially for fertiliser spreaders and crop sprayers. There are two basic types of plastic:

- Polythene and PVC are two examples of *thermoplastic plastic*. This material will become soft and pliable if subjected to gentle heat. Excessive heat will distort thermoplastics. They are used to make plastic sheeting, water pipes and fittings and buckets.
- *Thermosetting plastic* will not soften or melt when subjected to gentle heat, after it has hardened during manufacture. Like all plastics, it will be destroyed if exposed to high temperatures. Bakelite is one example of thermosetting plastic. It is used for electric plugs and switches, coil ignition distributor caps and other electrical components.

Fibreglass

Fibreglass is another material in this group. It is made of glass fibres held together with a special adhesive. It is liable to crack in service but can be mended with readily available repair kits. Fibreglass is used to make implement guards.

Rubber

Tyres, drive belts, engine hose pipes and flails on root harvesters are some examples of the use of rubber. This material will crack and perish if left in strong sunlight for long periods. Oil and grease will attack rubber and should be removed from tyres without delay.

Nuts and Bolts

Many types of nuts and bolts, together with various thread forms, can be found on the tractors and equipment on the farm.

Bolts

The hexagon-headed bolt is familiar to everybody and needs no description. Carriage bolts have a dome-shaped head with a square shoulder below. The square section can be used to grip in timber when it is bolted to a wooden or metal frame. The dome head is also used where a hexagon head might cause an obstruction; the square section holds the bolt while the nut is tightened. There is a variety of plough bolts in use too. These are used in a countersunk hole so that the bolt head does not protrude beyond the surface. The bolts used to hold a plough mouldboard in position are a good example. A small lug or a square section on the underside of the countersunk face stops the bolt from turning while the nut is tightened.

Machine screws may have round heads or countersunk heads. They are smaller in diameter than ordinary bolts and have thread right up to the head. There is a screwdriver slot or crosspoint recess in the head. Some countersunk machine screws have a hexagonal recess and are turned with an Allen socket. This is an L-shaped hexagonal section key or spanner (see Figure 21.3).

Set screws are used in a threaded hole and hold a component in position, usually on a shaft. They may have hexagon, square or slotted heads. Others have a hexagonal recess and are tightened with an Allen key. Set screws are usually threaded up to the head.

Nuts

The hexagon nut is the most common type of nut. A few square nuts are still in use. Wing nuts can be tightened with the fingers and are useful where frequent adjustments are required. There are different nuts in use which lock firmly on to a bolt and will not loosen through vibration. The castle nut has slots at the top; two of the slots in opposite faces of the nut are lined up with a hole through the bolt. A split pin is placed through the slots and the hole and the nut locked in position.

Lock nuts can also be used. These have a soft metal or plastic insert built into the nut. The hole in the insert is slightly smaller than the bolt thread. As the nut is tightened, the bolt cuts a thread in the insert. This is sufficient to lock the nut in position. This type of lock nut should only be used once if complete locking efficiency is required.

A nut can also be held tight on a bolt with a spring washer, a shake proof (or star) washer

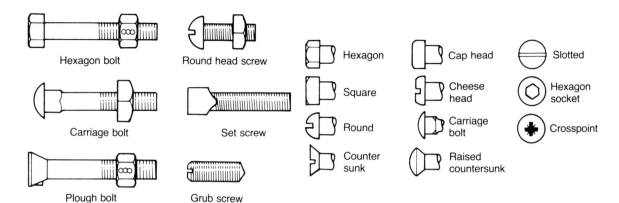

FIGURE 21.1 Nuts and bolts.

or by using two nuts, one tightened against the other, on the same bolt. Locking tabs on a tab washer can be bent upwards around the faces of a nut after it has been tightened. A tab washer is like an ordinary flat washer with two or more metal tabs around the circumference. It is usual to bend some of the tabs upwards and the remainder downwards over the component held in place by the nut.

Thread types

Nuts and bolts may have coarse or fine threads. A coarse thread has more strength than a fine one. The advantage of a fine thread is that it is less likely to vibrate loose in service. There are a number of different threads in common use and in almost every case a different set of spanners is required.

British Standard Whitworth (BSW) and British Standard Fine (BSF) are the coarse and fine threads used on much of the older British-made machinery. Spanners for these nuts and bolts are marked with the bolt diameter. A spanner which fits a ³/₈ in Whitworth bolt will have jaws much wider than ³/₈ in. A spanner which fits a ³/₈ in BSW nut will also fit a ¹/₁₆ BSF nut.

American National Coarse (ANC) and American National Fine (ANF) are used on machinery of American origin. In this case, the sizes on the spanners refer to the width across the flats of the nut. The spanners are marked A/F. A ³/₈ in A/F spanner fits a nut which is ³/₈ in across the flats.

Unified National Coarse (UNC) and Unified National Fine (UNF) are more recent threads. The A/F spanners will also fit unified nuts and bolts.

Metric threads have been used on continental-made machinery for many years and have now become standard on most British machinery. Spanners for metric nuts and bolts have the width across the flats of the nut marked on them. A metric spanner marked 22 will fit a nut which is 22 mm across the flats of the nut.

British Association thread (BA) is a fine rate thread for very small diameter nuts and bolts, used mainly for electrical equipment. The spanners are marked with a BA size and the most common sizes range from 0 BA (6 mm diameter) to 10 BA (1.7 mm diameter).

WORKSHOP TOOLS

Spanners

Open-ended spanners are one of the basic workshop tools. They are not always convenient for getting at nuts in confined spaces.

Ring spanners fit around the nut or bolt head and are less likely to slip than open-ended spanners. They are more suitable for getting at nuts in confined spaces. The ring spanner may have cranked ends or may be flat along its length.

Combination spanners offer the best of both types with an open-end jaw and a ring spanner combined. There is a different size at each end of ring and open-end spanners. This means that a set of either type has two spanners to fit the same size nut. Two sets of combination spanners would be needed to have the same size for both nut and bolt head.

Socket spanners are more expensive than the other types but have many advantages. With a ratchet they can be used at high speed. The socket can be used to get at nuts and bolts in recessed holes. A range of handles including speed brace, tee-handle and ratchet can be used with different extension bars to get at nuts inside components and all sorts of awkward places. A small universal coupling can be used so that the handle can be turned while at an angle to the socket. The standard range of sockets has a 12 mm square drive. Sockets with larger or smaller square drive can be obtained too.

Box spanners, more commonly known as plug spanners, are a cheap but effective form of socket spanner. It has six faces compared with

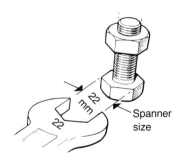

FIGURE 21.2 Spanner sizes.

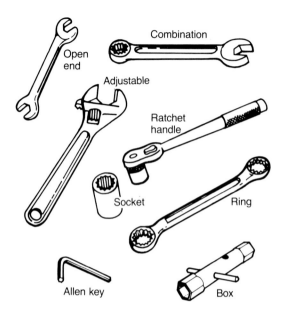

FIGURE 21.3　Types of spanner.

twelve on a socket spanner, and is turned with a steel rod (tommy bar).

Adjustable spanners are popular with all mechanics. A worn adjustable spanner with a stretched jaw is a dangerous tool. It will slip off nuts which are very tight, taking the corners off the nut and sometimes the skin off the user's knuckles. There are several sizes of adjustable spanner.

Allen keys are hexagonal sectioned and are 'L' shaped. They are used with Allen screws which have a recessed hexagonal head.

Torque spanners are used with sockets to tighten nuts or bolts to the required tightness (torque). This is a tool required by a mechanic when, for example, overhauling an engine.

Screwdrivers

The cabinet makers' screwdriver has a wooden handle and is meant for carpentry. The wooden handle must not be hit with a hammer.

Engineers' screwdrivers have round or square metal shafts in tough plastic handles. It is sometimes desirable to give the handle of the screwdriver a sharp tap when trying to undo a rusted screw. An engineers' screwdriver is suitable for this type of work.

Electricians' screwdrivers have insulated handles made of tough bakelite or plastic. They are used for work with mains electrical installations (with the supply turned off), vehicle electrics and light engineering work.

Phillips and Posidrive (cross-point) screwdrivers are designed for turning screws and small bolts with star-shaped recesses. Four point sizes are normally available.

Sharpening screwdrivers

This is a job best done with a file. A grindstone can be used but the tip of the blade must not be overheated as this will make it very brittle. The end of the blade should be flat and square and must not be sharpened to a knife edge. A screwdriver blade with rounded corners will soon spoil the slot in a screw head.

Cold Chisels

The general purpose cold chisel is used for trimming rough edges off metal, splitting seized nuts and a variety of other jobs. In time the handle end of a chisel will become mushroom shaped. In this condition it is dangerous, because a glancing

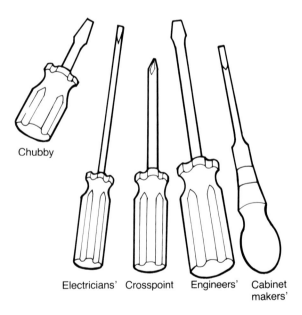

FIGURE 21.4　Screwdrivers.

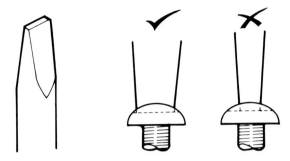

FIGURE 21.5 Sharpening screwdrivers.

blow with a hammer can cause part of the metal to break off. These chips of metal could injure you or other workers and the hammer could slip off the chisel with painful consequences. For safety's sake, grind off any burrs which form on the end of a chisel.

Sharpening chisels

Some cold chisels can be sharpened with a file, but a grindstone is normally used. Care must be taken to prevent overheating of the cutting edge. Frequent dipping in cold water will keep the chisel cool. A cold chisel is tapered towards the cutting edge, which should be sharpened at an angle of about 60 degrees.

Pliers

Engineers' or combination pliers can be found in all toolkits. The jaws have a flat face for holding flat material and a rounded section for gripping bolts and other round section material. The inner parts of the jaws have knife edges suitable

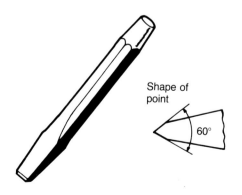

Shape of point

60°

FIGURE 21.6 Cold chisel.

for cutting wire or split pins. Some combination pliers also have slots on the outside of the jaws, near the hinge, which can be used for wire cutting.

Side-cutting pliers are designed for cutting wire, split pins and even nails. They are also very useful for removing stubborn split pins when dismantling parts of a tractor or machine.

Pointed nose pliers have rather limited use. They are useful for holding small parts in confined spaces.

Circlip pliers are specially designed for removing circlips from shafts, bearings and other components. Pliers for removing both internal and external circlips are available.

Files

Files are made in a variety of sizes, sections and types of cut. Single-cut files have rows of teeth running at an angle across the surface in one direction only. Double-cut files have rows of teeth running diagonally across the face of the file in two directions so that the teeth are shaped rather like small pyramids. Single-cut files are most suitable for cutting hard materials such as the knife sections of a cutter bar. Single-cut files are sometimes called reaper files. Double-cut files are used for most general filing work.

The number of teeth on a file varies considerably. This gives a range of files from smooth to rough. The following table gives an indication of the main file cuts:

Type	Teeth per cm	Teeth per inch
Smooth	24–26	60–65
Second cut	12–16	30–40
Bastard	8–12	20–30
Rough	6–8	15–20

A number of different file shapes or sections are made. The common file shapes include flat, square, round, half round, triangular (or three square) and flat with a safe edge. A safe-edge file has one smooth edge and is used to file out a corner without undercutting or removing metal with the file edge, while working with the face of the file.

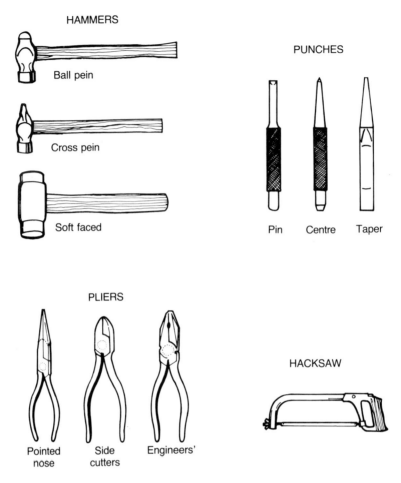

FIGURE 21.7   Hand tools.

Using files
File teeth are meant to cut on the forward stroke only. So take the pressure off the file on the backward stroke. Use the full length of the file and do not work too fast. About sixty strokes a minute is about right. Do not file dirty or rusty material as the teeth will soon become choked with dirt. A file should be cleaned with a file card. This is similar to a wire brush but has short wire bristles on a flexible backing.

Never store files in a way which allows them to rub against one another. Wrap them separately in a cloth or keep them upright in a stand. Always put a handle on a file before use. The sharp point, or tang, can cause injury to the hand. Some files have a shaped metal handle.

Hammers

The engineers' or ball pein hammer is the most useful type of hammer for the farm workshop. It has a ball at one end of the head and a flat face on the other. The ball pein is used for riveting.

Cross pein hammers have a chisel-shaped end opposite the flat face. A claw hammer has a claw for removing nails. The cross pein and claw hammer are mainly for the carpenter.

The club hammer (or lump hammer), used by bricklayers, is a handy farm workshop tool.

The sledge hammer is a necessary farm workshop tool. It is ideal for straightening damaged or bent components, especially if an anvil is available.

Hammers are made in a range of weights. A small 0.23 kg (½ lb) hammer is very good for cutting gaskets. An engineers' hammer of about 1.1 kg (2½ lb) is a good general purpose tool. A 3.2 or 4.5 kg (7 or 10 lb) sledge hammer is another useful piece of equipment.

Using hammers

Always hold a hammer close to the end of the handle or shaft. Never use a hammer with a loose head or damaged handle. The head can be tightened by driving an extra wedge into the handle where it is attached to the head. The faces of hammers are very hard; do not hit one hammer against another as this can damage the faces.

Soft-faced hammers should be used where an ordinary hammer will damage the component. This is the case with brittle or soft materials such as cast iron, copper and brass. Soft-faced hammers are usually made of copper, hide or rubber.

Hacksaws

Most hacksaw frames are adjustable for different lengths of blade; 25 or 30 cm (10 or 12 in) blades are in common use. A junior hacksaw with much shorter blades is also made.

The blade must be fitted with the teeth pointing away from the handle. To tension a hacksaw blade, take up the slack and then tighten a further two or three turns. The number of teeth on a blade must suit the job in hand. A coarse blade with 7 teeth per cm (18 teeth per in) is best for thick or soft material. A blade with 12.6 teeth per cm (32 teeth per in) is needed for cutting very thin metal. For general work, a 9.4 teeth per cm (24 teeth per in) is best. The important thing to remember is to make sure that at least two teeth are always in contact with the metal.

There is a choice between flexible or hard blades. The flexible blade is best for the handyman. It will not break when used carelessly. The hard blade gives more accurate results but is rather brittle and will snap if the saw is twisted while in use.

Using hacksaws

Always use the full length of the blade. Use a steady sawing speed of about sixty strokes per minute. Exert a steady pressure on the saw during the forward stroke. Release the pressure when pulling the saw towards you. Very little pressure is needed when cutting thin section metal. Make sure that the work is secure in a vice before using the hacksaw.

Punches

Taper punches are made in various sizes and are mainly used to line up bolt or pin holes when joining two components with rivets, bolts, etc.

Parallel, or pin, punches do not taper to a narrow point. They are used to remove bolts and pins from such components as pulleys, gears and shafts. Pin punches may break if used to align holes as they are not designed to act as levers.

Centre punches are used to mark the centre spot in preparation for drilling a hole in metal. The punch mark also provides a guide for the point of the drill.

Rivet snaps are a special type of punch. The end of the punch has a hole shaped like a rivet head. After swelling the rivet in its hole with a hammer, a rivet snap can be used to form a neat finish to the work.

Drills

Hand and power drills are found in most workshops. A set of twist drills for working with metal may be suitable for either high or low speed work. High carbon steel drills are relatively cheap but will not last for very long if used in a high speed electric drill. Twist drills for high speed work are made from steel with a high tungsten content. They are more expensive but with care will give long service. Twist drills for hand-held drills have parallel shanks. The shank must be tight in the drill chuck. Drill sharpening is a skilled job. The point should have an angle of 60 degrees. Sharpen a twist drill with a grinder but avoid overheating the point. Use a new twist drill as a guide to get the correct angles if you cannot get someone to show you how to sharpen a drill.

Taps and Dies

These are used for cutting threads or restoring damaged threads. A tap is used to cut a thread in a hole and a die is used to make a thread on a piece of rod. Die nuts, which are turned with a spanner, are useful for cleaning rusty or damaged threads.

POWER TOOLS

Compressors

Most air compressors are driven by an electric motor. They provide compressed air for tyres, spray painting and cleaning equipment. Compressed air can also be used to operate power tools such as drills and grinders without the risk of an electric shock.

Never direct air pressure guns at the body as this can cause serious injury.

Drills

Most electric power tools run at mains voltage (240 volts). Some heavy duty equipment is connected to 415 volt three-phase supply. It is possible to buy electrical tools which run at lower voltages for added safety. Many manufacturers offer, for example, drills at 240 or 110 volts.

Drills are essential workshop tools. With suitable attachments, they can be used for sawing, sanding, wire brushing and polishing. There are many sizes from the handyman's 8 mm capacity drill up to a heavy duty 19 mm capacity model. Larger drills either fixed in drill stands or pedestal models can be used to drill holes of 38 mm diameter or more.

Using drills

Clamp the work securely, especially when using a stand-mounted or pedestal drill. Do not force the drill through the metal; if it will not cut it is probably blunt. Large holes can be drilled more efficiently if a small (pilot) hole is drilled first.

A better finish is obtained if the drill has proper lubrication. Steel should be lubricated with water-soluble oil. This is a white fluid which is mixed with water. It is often known as cutting oil. It has good lubricating and cooling properties for drilling steel.

Cast iron should be drilled dry. Soft metals such as copper, brass and aluminium can be drilled dry or with a lubricant such as paraffin.

When drilling a hole in a piece of sheet metal, better results can be obtained by resting the metal on a piece of wood.

Grinders

Many types of grinder from small bench models to heavy duty floor standing models are available. Grinders have two wheels; it is usual to have one coarse and one fine grinding wheel. The fine wheel should be reserved for sharpening drills, knives and similar precision items. A wire brush can be fitted in place of a grinding wheel if required.

Using grinders

Try to use the full face of the grinding wheel. Never exert any pressure when using the side of the wheel.

Do use goggles. This is the only way to be sure of keeping stray metal chips and sparks out of your eyes. The rests around the grinding wheels can be adjusted. Make sure that the clearance between the rest and the grinding wheel is very small. The part being ground can be pulled between the wheel and the rest if the gap is too wide.

In time the grinding wheels may become uneven. The wheel faces can be restored to the correct shape with a stone dresser. One type has some very hard rotary cutters which are held against the face of the grinding wheel until a true edge is produced.

Angle Grinders

These are similar to an electric drill but instead of a chuck for a twist drill, there is a grinding disc at right angles to the motor drive shaft. The disc can be used to grind parts of machinery or very large pieces of metal. The grinder is taken to the work rather than taking the work to the grinder. The grinding disc can be replaced with a cutting disc which can cut a wide range of materials. These include metal, masonry and roof sheeting.

Take great care when using an angle grinder. Make sure the cable is safely out of the way. Do not let sparks fly towards other people and remember to wear grinding goggles.

SAFETY IN THE FARM WORKSHOP

Poor working conditions, unsafe hand tools, faulty electrical equipment, untidy benches and dirty floors can cause accidents in the workshop.

A few simple precautions combined with general cleanliness will go a long way towards preventing accidents.

The workshop

A warm, well-lit building with a concrete floor is a good basis for establishing a farm workshop. It is so easy to bruise or cut hands which are cold while doing maintenance work in the winter time.

Hand tools

Periodic checks of hand tools can prevent accidents. Look for loose hammer heads, mushroom-headed chisels, open-ended spanners with stretched jaws and worn adjustable spanners. Where the faults cannot be put right, the tool should be thrown into the scrap bin. Always put a handle on a file before you use it.

Do not wrap your fingers tightly round a spanner. Exert the pressure on the spanner with the palm of the hand when first undoing a nut, then if the spanner slips, there is far less risk of losing the skin off your knuckles.

Electrical equipment

Power-operated tools such as drills and grinders are a source of danger if they are not well maintained. Never use an electric tool unless it is properly earthed or double insulated. This is particularly important when using the equipment outside in a farmyard where the cable trails over damp ground or concrete.

Check the condition of all cables and plugs frequently. Any damage must be reported at once. Electricity can kill, so treat it with great respect. When a fuse blows always use a fuse of the same size and rating to replace it.

Jacks

Never work under a jacked-up vehicle. Always support it with axle stands or wooden blocks before going underneath to carry out repairs. It is good practice to place chocks under any wheels left on the ground. Avoid leaving a vehicle jacked up without any other means of support—someone may accidentally let it down. Always keep jacks in good condition.

The first aid box

To comply with the Farm Safety Regulations, all farms must have a first aid box. Its contents will vary with the number of employees on the farm. You can obtain a leaflet from the local Farm Safety Officer, which lists all the bandages, ointments, etc. which must be in the box. A copy of the official farm safety leaflet about first aid must also be in the box. At least one member of the farm staff should have basic training in first aid.

Asbestos

This material was widely used in the past around the farm but has now been proved a health hazard. Other material, including metal, plastic and non-asbestos cement sheeting and rainwater goods should be used instead.

If asbestos must be handled, wear full protective clothing; however, it is strongly recommended that expert advice be sought. Better still, a specialist contractor should be employed to remove asbestos safely, thus preventing any risk to the health of the farm staff.

Note: some clutch and brake linings have an asbestos content. Do not use an air line to clean the dust away if working on these components. Wear a mask to avoid inhaling any dust.

Personal Protection

Always wear goggles when using a grinder or portable electric tool which causes sparks. Wear substantial footwear. Shoes will give no protection against heavy objects falling on your feet. Safety boots with steel toe-caps are ideal for the workshop. Heavily studded boots can be dangerous on oily floors or when climbing over metal surfaces on machines.

Wash your hands thoroughly, especially after working with spraying machinery. Many people are allergic to diesel fuel and they may suffer from a skin complaint called dermatitis. Such troubles can be prevented by using barrier cream before starting work.

The best clothing for the workshop is a boiler suit. Long flowing clothes can involve the wearer in serious accidents, especially when working with moving machinery.

Fire

Store all fuel and oil away from the workshop if possible and in any case well away from any welding or cutting equipment. Never oil the threads on oxyacetylene welding gauges and hose connections. The oil can cause an explosion.

Keep all fire extinguishers easily accessible. It is most important to know which type of extinguisher to use on a particular fire. An extinguisher which contains liquid must never be used on a fire caused by an electrical fault. Switch off the electricity supply to the building before tackling the fire. Make sure that the fire brigade has been called.

There are several types of fire extinguisher. Make sure you know the different types and when it is safe to use them.

Dry powder and carbon dioxide extinguishers can be used against live electrical fires and burning liquids. They are only suitable for small fires involving wood, paper and similar materials.

Foam extinguishers must not be used on live electrical fires because the foam is a conductor of electricity. This type can be used to tackle burning liquids and small fires where wood, paper or similar material are involved.

Soda acid extinguishers are water based and must not be used to fight live electrical fires or burning liquids such as oil, paint or grease. Soda acid extinguishers are best suited to tackle burning materials such as paper, wood and other materials.

Gas pressure extinguishers are also water based. When gas is released inside the extinguisher it forces out a jet of water. Like the soda acid extinguisher, it is not suitable for burning liquids and live electrical fires.

As well as fire extinguishers, keep some buckets of sand, a fire blanket and some buckets of water in the workshop and other high risk areas.

Suggested Student Activities

1. Identify the different types of metal you may find in a farm workshop.
2. Learn the different spanner sizes. You will be proficient when you can pick the correct spanner for the common bolt sizes at the first attempt.
3. Look for the range of spanners and other tools listed in this chapter. Learn their names.
4. Practise using a hacksaw and a file. Use all the teeth on the blade or file and try to keep the work square.
5. Ask someone to show you how to sharpen a twist drill.

Safety Check

Eyes and ears are at particular risk in the workshop. Wear ear defenders when using a noisy power tool such as an angle grinder. Always use goggles to protect your eyes when grinding or blowing dirt from machinery. Always wear the correct eye protection equipment when using or watching a welder.

METRIC CONVERSION TABLES

BRITISH TO METRIC

Length

1 inch (in)	= 2.54 cm
	or 25.4 mm
1 foot (ft)	= 0.30 m
1 yard (yd)	= 0.91 m
1 mile	= 1.61 km

Conversion factors

inches to cm	× 2.54
or mm	× 25.4
feet to m	× 0.305
yards to m	× 0.914
miles to km	× 1.61

Area

1 sq inch (in^2)	= 6.45 cm^2
1 sq foot (ft^2)	= 0.093 m^2
1 sq yard (yd^2)	= 0.836 m^2
1 acre (ac)	= 4047 m^2
	or 0.405 ha

Conversion factors

sq feet to m^2	× 0.093
sq yards to m^2	× 0.836
acres to ha	× 0.405

Volume (Liquid)

1 fluid ounce (1 fl oz)	
(0.05 pint)	= 28.4 ml
1 pint	= 0.568 litres
1 gallon (gal)	= 4.55 litres

Conversion factors

Pints to litres	× 0.568
gallons to litres	× 4.55

METRIC TO BRITISH

Length

1 millimetre (mm)	= 0.0394 in
1 centimetre (cm)	= 0.394 in
1 metre (m)	= 1.09 yd
1 kilometre (km)	= 0.621 miles

centimetres to in	× 0.394
millimetres to in	× 0.0394
metres to ft	× 3.29
metres to yd	× 1.09
kilometres to miles	× 0.621

Area

1 sq centimetre (cm^2)	= 0.16 in^2
1 sq metre (m^2)	= 1.20 yd^2
1 sq metre (m^2)	= 10.8 ft^2
1 hectare (ha)	= 2.47 ac

sq metres to ft^2	× 10.8
sq metres to yd^2	× 1.20
hectares to ac	× 2.47

Volume (Liquid)

100 millilitres (ml or cc)	= 0.176 pints
1 litre	= 1.76 pints
1 kilolitre (1000 litres)	= 220 gal

litres to pints	× 1.76
litres to gallons	× 0.220

Weight

1 ounce (oz)	= 28.3 g	1 gram (g)	= 0.053 oz
1 pound (lb)	= 454 g	100 grams	= 3.53 oz
	or 0.454 kg	1 kilogram (kg)	= 2.20 lb
1 hundredweight (cwt)	= 50.8 kg	1 tonne (t)	= 2204 lb
1 ton	= 1016 kg		or 0.984 ton
	or 1.016 t		

Conversion factors

ounces to g	× 28.3	grams to oz	× 0.0353
pounds to g	× 454	grams to lb	× 0.00220
pounds to kg	× 0.454	kilograms to lb	× 2.20
hundredweights to kg	× 50.8	kilograms to cwt	× 0.020
hundredweights to t	× 0.0508	tonnes to tons	× 0.984
tons to kg	× 1016.0		
tons to tonnes	× 1.016		

INDEX

INDEX

FARMING PRESS BOOKS

Below is a sample of the wide range of agricultural and veterinary books published by Farming Press. For more information or for a free illustrated book list please contact:

**Farming Press Books, 4 Friars Courtyard
30–32 Princes Street, Ipswich IP1 1RJ, United Kingdom
Telephone (0473) 43011**

ALSO BY BRIAN BELL ● **Farm Workshop**

Describes the basic requirements of the farm workshop and illustrates the uses of the necessary tools and equipment. Techniques covered range from simple metalwork to electric arc welding and tractor maintenance.

In the same Farm Workshop Series (paperback)

GRAHAM BOATFIELD ● **Farm Livestock**

An excellent introduction to livestock production methods designed for those taking their first steps in agriculture. Full details on cattle, sheep and pigs, and general sections on feeding, breeding and health.

GRAHAM BOATFIELD ● **Farm Crops**

Gives an outline of farm crops and crop husbandry, along with basic knowledge and scientific principles underlying the growing of those crops. Ideal for students on introductory farming and rural science courses.

GRAHAM BOATFIELD AND IAN HAMILTON ● **Calculations for Agriculture and Horticulture**

Provides the calculation methods for crops, livestock, machinery and horticulture. Ideal for young people entering agricultural training or work.

F. RAYMOND, P. REDMAN AND R. WALTHAM
Forage Conservation and Feeding
Includes all the latest information on hay and silage making, mowing and field treatments, grass drying and forage feeding.

D. B. DAVIES, D. J. EAGLE AND B. FINNEY
Soil Management
Goes into all the aspects of the soil, plant nutrition, farm implements and their effects on the soil, crop performance, land drainage and cultivation systems.

MAURICE BARNES AND CLIVE MANDER
Farming Building Construction
Covers practical farm building work. Details on blockwork, brickwork, timber, flooring, roads, walls, etc., for new constructions or improvements.

JOHN ARCHER
Crop Nutrition and Fertiliser Use
Gives details of uptake for each nutrient and then deals with the specific requirements of temperate crops from grassland and cereals to vegetables, fruit and nursery stock.

Farming Press also publish three monthly magazines: *Dairy Farmer, Pig Farming* and *Arable Farming*. For a specimen copy of any of these magazines, please contact Farming Press at the address above.